AGRICULTURE ISSUES AND POLICIES

NITROGEN USE IN UNITED STATES AGRICULTURE

IMPLICATIONS AND MANAGEMENT

AGRICULTURE ISSUES AND POLICIES

Additional books in this series can be found on Nova's website
under the Series tab.

Additional E-books in this series can be found on Nova's website
under the E-book tab.

ENVIRONMENTAL REMEDIATION TECHNOLOGIES, REGULATIONS AND SAFETY

Additional books in this series can be found on Nova's website
under the Series tab.

Additional E-books in this series can be found on Nova's website
under the E-book tab.

AGRICULTURE ISSUES AND POLICIES

NITROGEN USE IN UNITED STATES AGRICULTURE

IMPLICATIONS AND MANAGEMENT

EVERETT F. WILKINSON
AND
EDSON J. PARKER
EDITORS

Nova Science Publishers, Inc.
New York

For permission to use material from this book please contact us:
Telephone 631-231-7269; Fax 631-231-8175
Web Site: http://www.novapublishers.com

Library of Congress Cataloging-in-Publication Data

ISBN: 978-1-62081-636-3

Published by Nova Science Publishers, Inc. † New York

CONTENTS

PREFACE

Nitrogen is an important agricultural input that is critical for crop production. However, the introduction of large amounts of nitrogen into the environment has a number of undesirable impacts on water, terrestrial, and atmospheric resources. This book explores the use of nitrogen in U.S. agriculture and assesses changes in nutrient management by farmers that may improve nitrogen use efficiency. It also reviews a number of policy approaches for improving nitrogen management and identifies issues affecting their potential performance.

Chapter 1 - Nitrogen is an important agricultural input that is critical for crop production. However, the introduction of large amounts of nitrogen into the environment has a number of undesirable impacts on water, terrestrial, and atmospheric resources. This report explores the use of nitrogen in U.S. agriculture and assesses changes in nutrient management by farmers that may improve nitrogen use efficiency. It also reviews a number of policy approaches for improving nitrogen management and identifies issues affecting their potential performance. Findings reveal that about two-thirds of U.S. cropland is not meeting three criteria for good nitrogen management. Several policy approaches, including financial incentives, nitrogen management as a condition of farm program eligibility, and regulation, could induce farmers to improve their nitrogen management and reduce nitrogen losses to the environment.

Chapter 2 - Policymakers are dedicating considerable attention to greenhouse gas emission reduction, primarily discussing options for carbon dioxide (CO_2) emission reduction.[1] Less frequently addressed in proposed legislation is emission reduction for non- CO_2 greenhouse gases, such as nitrous oxide (N_2O). However, N_2O reduction efforts have the potential to mitigate climate change. Moreover, N_2O emission sources may be regulated under the existing Clean Air Act as a class I or class II ozone-depleting substance at the discretion of the Environmental Protection Agency (EPA) Administrator. No new legislation needs to be passed to regulate N_2O for climate protection and ozone recovery.

In: Nitrogen Use in United States Agriculture ISBN: 978-1-62081-636-3
Editors: E.F. Wilkinson, E.J. Parker

Chapter 1

NITROGEN IN AGRICULTURAL SYSTEMS: IMPLICATIONS FOR CONSERVATION POLICY*

Marc Ribaudo, Jorge Delgado, LeRoy Hansen, Michael Livingston, Roberto Mosheim and James Williamson

ABSTRACT

Nitrogen is an important agricultural input that is critical for crop production. However, the introduction of large amounts of nitrogen into the environment has a number of undesirable impacts on water, terrestrial, and atmospheric resources. This report explores the use of nitrogen in U.S. agriculture and assesses changes in nutrient management by farmers that may improve nitrogen use efficiency. It also reviews a number of policy approaches for improving nitrogen management and identifies issues affecting their potential performance. Findings reveal that about two-thirds of U.S. cropland is not meeting three criteria for good nitrogen management. Several policy approaches, including financial incentives, nitrogen management as a condition of farm program

* This is an edited, reformatted and augmented version of United States Department of Agriculture, Economic Research Service, Economic Research Report No. 127, dated September 2011.

eligibility, and regulation, could induce farmers to improve their nitrogen management and reduce nitrogen losses to the environment.

ACKNOWLEDGMENTS

This report benefited from the insightful comments provided by Keith Fuglie, Marca Weinberg, and Mary Bohman of USDA's Economic Research Service, Ralph Heimlich of Agricultural Conservation Economics, Roberta Parry of the U.S. Environmental Protection Agency, Jerry Hatfield of USDA's Agricultural Research Service, USDA's Natural Resources Conservation Service and an anonymous reviewer. Thanks also to John Weber for excellent editorial assistance and to Curtia Taylor for the design and layout.

SUMMARY

What Is the Issue?

Nitrogen is an agricultural input that is critical for crop production. Human-induced production and release of reactive nitrogen has greatly affected the Earth's natural balance of nitrogen, contributing to changes in ecosystems, both beneficial and harmful, including increased agricultural productivity in nitrogen-limited areas, ozone-induced injury to crops and forests, over-enrichment of aquatic ecosystems, biodiversity losses, visibility-impairing haze, and global climate change. Incentives for encouraging farmers to adopt improved nitrogen management can take many forms, from purely voluntary to regulatory. Designing a cost-effective policy requires that factors influencing fertilizer use be fully understood. Also, an understanding of how farmers are likely to respond to different incentives may help policymakers assess potential environmental tradeoffs driven by nitrogen's ability to change forms and cycle through different environmental media.

What Did the Study Find?

- Emission of reactive nitrogen to the environment can be reduced by matching nitrogen applications more closely with the needs of growing crops. This can be achieved by adopting three "best management practices" (BMPs):

- *Rate*: Applying an amount of nitrogen at a rate that accounts for all other sources of nitrogen, carryover from previous crops, irrigation water, and atmospheric deposits.
- *Timing:* Applying nitrogen as close to the time that the crop needs it as is practical (as opposed to the season before the crop is planted).
- *Method:* Injecting or incorporating the nutrients into the soil to reduce runoff and losses to the atmosphere.

- Among all U.S. field crops planted in 2006 that received nitrogen fertilizers, 35 percent are estimated to have met all three of the nutrient BMPs. For the remaining cropland, improvements in management are needed to increase nitrogen use efficiency (i.e., reduce the amount of nitrogen available for loss to the environment).
- Corn is the most intensive user of nitrogen fertilizer, on a per acre basis and in total use. Fertilizer applied to corn is least likely to be applied in accordance with all three BMPs.
- Incentives for improving nitrogen use efficiency by adopting the rate, timing, and method BMPs can come from policy or market forces:
    - Government programs that provide financial assistance for adopting BMPs can be effective if they encourage the participation of farmers with land most in need of improvement and if the farmers choose the most cost-effective practices. Data suggest that the amount of cropland needing improvement would require a substantial increase in the current Federal budget devoted to nutrient management practices.
    - Including nitrogen management in compliance provisions for receiving Federal farm payments could encourage farmers to adopt more effective management practices. In 2005, producers of U.S. corn received government payments that were much higher than the cost of improving nitrogen management. The strength of this incentive, however, has declined in recent years because of increases in crop prices and a decline in direct commodity payments.
    - Emissions markets, such as water quality trading and greenhouse gas cap-and-trade, could provide financial incentives to farmers to adopt improved nitrogen management and produce nitrogen credits that can be sold in these markets. The effectiveness of

such markets would depend on market design, including rules defining who can participate and what needs to be done to produce credits.

- Onfield improvements to nitrogen use efficiency could be supplemented with off-field practices, such as wetlands restoration and vegetative filter strips that can filter and trap reactive nitrogen that leaves the field through surface runoff and groundwater flow. Of the two practices, restored wetlands can be more cost effective at removing nitrogen and provide additional environmental benefits, but they are limited to areas with suitable soils and hydrology. Vegetative filters can be employed more widely across the landscape but are not effective when existing tile drains bypass the filters.
- Policies for increasing nitrogen use efficiency should recognize the potential environmental tradeoffs when addressing particular issues related to reactive nitrogen. Focusing strictly on one issue, such as nitrate leaching, could lead to increased emissions of other nitrogen compounds, such as nitrous oxide, even when total reactive nitrogen emissions are reduced.

How Was the Study Conducted?

ERS researchers used an extensive literature review, modeling, and data from USDA's Agricultural Resource Management Survey (ARMS) of major field crops. ARMS data provided information on nitrogen use, defined by the rate, method, and timing application criteria. This, in turn, helped researchers determine the types of management improvements needed the most.

The following market forces and policy instruments were evaluated to measure their influence on nitrogen management: nitrogen fertilizer taxes, Federal financial assistance offered to farmers to adopt practices that improve nitrogen use efficiency or filter and trap nitrogen runoff, emissions markets such as water quality trading and greenhouse gas cap-and-trade, compliance with nitrogen BMPs as a condition for receiving farm program benefits, and regulation.

Because reactive nitrogen is mobile and able to transform into different compounds, researchers used a field-level nitrogen loss simulator developed by USDA's Agricultural Research Service to track how improving nitrogen use efficiency by meeting all three BMPs affects emissions of different reactive nitrogen compounds. These interactions were taken into account when evaluating alternative policy options.

GLOSSARY

ARMS	Agricultural Resource Management Survey
BMP	Best management practice
CEAP	Conservation Effects Assessment Program
EQIP	Environmental Quality Incentives Program
NUE	Nitrogen use efficiency
N	Nitrogen
N_2	Gaseous nitrogen
NO_3	Nitrate
NO_x	Nitrogen oxides
N_2O	Nitrous oxide
NH_3	Ammonia
Nr	Reactive nitrogen
NRCS	Natural Resources Conservation Service (USDA)
VFS	Vegetative filter strip

INTRODUCTION

Most of the cropping systems in the world are naturally deficient in nitrogen, making nitrogen inputs necessary to produce the crop yields needed to support human populations. Gaseous nitrogen (N_2) is abundant in the atmosphere, but it cannot be used by living organisms unless it is first converted into useable forms. Leguminous plants and soil microorganisms contribute significant amounts of nitrogen used by crops, but yields necessary to support growing populations need more nitrogen than can be provided by natural means.

The Haber-Bosch process converts "unreactive" gaseous nitrogen from the atmosphere into a biologically useable "reactive" form. The development of the process in the early 1900s led to the massive production of relatively inexpensive nitrogen fertilizer that boosted crop yields (Follett et al., 2010). The increasing use of reactive nitrogen in agriculture also increased the potential for nitrogen to be lost to the environment as ammonia (NH_3), ammonium

(NH_4), nitrogen oxides (NO_x), nitrous oxide (N_2O), and nitrate (NO_3); these compounds are all reactive forms of nitrogen (Galloway et al., 2003). Excessive amounts of reactive nitrogen inputs can lead to imbalances in the natural movement of nitrogen among atmospheric, terrestrial, and aquatic nitrogen pools, leading to disruptions in ecosystem function and the supply of valuable ecosystem services.

Reactive nitrogen directly affects species composition, diversity, dynamics, and the functioning of terrestrial, freshwater, and marine ecosystems (Matson et al., 1997; Vitousek et al., 1997). Human-induced increases in reactive nitrogen emissions to the environment may contribute to the following harmful changes to ecosystems:

- Ozone-induced injury to crop, forest, and natural ecosystems
- Acidification and eutrophication (nutrient enrichment) effects on forests, soils, and freshwater aquatic ecosystems
- Eutrophication and hypoxia (oxygen depletion) in coastal and lake ecosystems
- Harmful algae blooms
- Biodiversity losses in terrestrial and aquatic ecosystems
- Regional haze
- Depletion of stratospheric ozone
- Global climate change
- Nitrate contamination of drinking water aquifers

A variety of steps can be taken to reduce the relatively large share of nitrogen that is lost from agricultural systems. Improved management of nitrogen fertilizers, animal manure, and other agricultural inputs can improve overall nitrogen use efficiency (NUE) and reduce the loss of reactive nitrogen to the environment while maintaining crop yields.

Incentives for encouraging farmers to adopt improved nitrogen management can take many forms, from purely voluntary to regulatory. Designing a cost-effective policy requires that factors influencing fertilizer use be fully understood. Also, an understanding of how farmers are likely to respond to different incentives may help policymakers assess potential environmental tradeoffs driven by nitrogen's ability to change forms and cycle through different environmental media.

This report takes a broad view of several questions related to nitrogen management: (1) Why is nitrogen management so important? (2) How many acres of cropland are not using nitrogen best management practices (BMP)?

and (3) What are the strengths and weaknesses of alternative policy approaches for improving nitrogen management on those acres?

Ideally, alternative policies would be assessed on the basis of the cost of achieving a particular level of NUE across U.S. crop production. However, physio-economic models that would allow for this type of assessment are not available on a national scale. Instead, this analysis uses survey data to help identify the number of acres of cropland that would benefit from improved management and to assess the characteristics of each alternative policy approach. Policy approaches are assessed in terms of factors consistent with cost effectiveness, including flexibility, ability to target, crop acres covered, and implementation costs. These factors are assessed through original research and an extensive review of the literature.

Environmental Implications of Nitrogen and Goals for Agricultural Management

Agriculture is the predominant source of reactive nitrogen emissions into the environment. In the United States, agriculture contributes 73 percent of nitrous oxide emissions (EPA, 2010a), 84 percent of ammonia emissions (EPA, 2010a), and 54 percent of nitrate emissions (Smith et al., 1997). Most losses from cropland are attributable to runoff, ammonia volatilization, nitrification and denitrification, and nitrate leaching (see box, "Pathways for Nitrogen Losses").

Nitrogen's impacts on water resources (Dubrovsksy et al., 2010; Bricker et al., 2007; Rabalais et al., 2002a, b), atmosphere (Cowling et al., 2002; Follett et al., 2010), and terrestrial resources (Galloway et al., 2008) are extensive. Estimates of the economic value of these damages are lacking. Crutchfield et al. (1995) estimate that consumers in four U.S. rural areas would be willing to pay between $73 million and $780 million per year (in 1995 dollars) for reduced chemical concentrations (including nitrate) in groundwater tapped by private wells. Dodds et al. (2009) estimate that consumers spend over $800 million each year on bottled water due to nutrient-related taste and odor problems.

Using data from water treatment plants, ERS estimates the cost of removing nitrate from U.S. drinking water supplies is over $4.8 billion per year (see app. 1). Based on the contribution of nitrate loadings from agriculture (Smith et al., 1997), agriculture's share of these costs is estimated at about $1.7 billion per year. Most costs are borne by the large utilities, due to the volume of water treated. ERS findings indicate that reducing nitrate concentrations in source waters by 1 percent would reduce water treatment costs in the United States by over $120 million per year.

Pathways for Nitrogen Losses

Soil erosion - Nitrogen can be lost from the soil surface when attached to soil particles that are carried off the field by wind or water. Although wind and water erosion can be observed across all regions, wind erosion is more prevalent in dry regions and water erosion in humid regions. Overall, little nitrogen is lost through erosion when basic conservation practices are in place (Iowa Soybean Association, 2008b).

Runoff - Surface runoff of dissolved nitrogen (generally in the form of nitrate) is only a concern when fertilizer and or manure are applied on the surface and rain moves the nitrogen before it enters the soil (Legg and Meisinger, 1982; Iowa Soybean Association, 2008b).

Ammonia volatilization - Significant amounts of nitrogen can be lost to the atmosphere as ammonia (NH_3) if animal manure or urea is surface applied and not immediately incorporated into the soil (Hutchinson et al., 1982; Fox et al., 1996; Freney et al., 1981; Sharpe and Harper, 1995; Peoples et al., 1995). Additionally, warm weather conditions can accelerate the conversion of manure and other susceptible inorganic nitrogen fertilizers to ammonia gas.

Denitrification and nitrification - When oxygen levels in the soil are low, some microorganisms known as denitrifiers will convert NO_3 to nitrogen (N_2) and nitrous oxide (N_2O), both of which are gases lost to the atmosphere (Mosier and Klemedtsson, 1994). Nitrogen gas is not an environmental issue, but N_2O is a powerful greenhouse gas. Denitrification usually occurs when nitrate is present in the soil, soil moisture is high or there is standing water, and soils are warm. NO_x and N_2O gases can also be produced through a process called nitrification.

Leaching - Leaching occurs when there is sufficient rain and/or irrigation to move easily dissolvable nitrate through the soil profile (Randall et al., 2008). The nitrate eventually ends up in underground aquifers or in surface water via tile drains and groundwater flow. Tile drains may be a chief passageway by which nitrogen moves from crop soils to surface water (Turner and Rabalais, 2003; Randall et al., 2008; Randall et al., 2010).

Managing Nitrogen for Agriculture and the Environment

USDA's Natural Resources Conservation Service (NRCS) defines nutrient management as managing the amount, source, placement, form, and timing of the application of plant nutrients to the soil (USDA, NRCS, 2006). Optimizing nitrogen management both economically and environmentally requires farmers to perform a juggling act: Applying too much nitrogen cuts into financial returns and increases the likelihood of nitrogen escaping into the environment; applying too little increases the risk of reduced yields and lost income. Crop production is characterized by uncertain and stochastic, or random, weather and soil conditions that affect crop yields and nitrogen loss. To maintain viable operations, farmers may manage temporal variability in weather and soil nitrogen by overapplying nitrogen to protect against downside risk (i.e., use an "insurance" nitrogen application rate) (Sheriff, 2005; Babcock, 1992; Babcock and Blackmer, 1992). Additionally, farmers may take a "safety net" approach to maximize economic returns by setting an optimistic yield goal for a given field based on an optimum weather year to ensure that the needed amount of nitrogen for maximum yields is available (Schepers et al., 1986; Bock and Hergert, 1991). Thus, during the years in which weather is not optimal for maximizing yields, nitrogen will be overapplied from an agronomic standpoint. Almost by definition, optimal conditions are infrequent, so farmers overfertilize crops in most years. The following hypothetical example helps illustrate the reasoning behind a farmer's decision to apply a certain amount of fertilizer. Assume that a farmer applies 179 pounds of nitrogen (N) per acre to his or her cornfield. Under ideal conditions, the farmer might produce 170 bushels of corn per acre. In most years, however, conditions are not ideal and production averages 148 bushels per acre. This yield requires only 165 pounds of N per acre, but at this level, the farmer will miss out on an extra 22 bushels in the event of ideal weather conditions. Assuming a fertilizer price of $0.50 per pound of N, the extra N applied in an average year costs $7 per acre. Assuming a corn price of $4.50 per bushel, the benefit from having enough nitrogen available to take advantage of optimal conditions would be $99 per acre. In most years, however, the extra fertilizer is not used by the crop and is available to leave the field and affect environmental quality.

Definitions of Nitrogen Use Efficiency

Researchers calculate nitrogen use efficiency to assess the effectiveness of nitrogen management. The NUE of a cropping system is the proportion of all nitrogen inputs that are removed in harvested crop biomass, contained in recycled crop residues, and incorporated in soil organic and inorganic nitrogen pools (Cassman et al., 2002) (fig. 2.1). Nitrogen not recovered in these nitrogen sinks is lost to the environment. Increases in NUE reduce the share of nitrogen left in the soil and available for loss to water or the atmosphere. Increased NUE is treated as a goal of environmental policy throughout this report.

Recommended Input Rate and Nitrogen Credits

The nitrogen application rate has a major effect on NUE (Bock and Hergert, 1991; Meisinger et al., 2008; Freney et al., 1995; Power et al., 2001). Nitrogen losses have been shown to increase rapidly when N inputs exceed assimilation capacity (Vanotti and Bundy, 1994; Schlegel et al., 1996; Dobermann et al., 2006; Bock and Hergert, 1991). Reducing application rates reduces the losses of all forms of reactive nitrogen.

The effectiveness of nitrogen management may be raised by accounting for all nitrogen sources when determining a nitrogen fertilizer application rate. Depending on the region, such sources may include inorganic nitrogen levels in the root zone, soil organic content, previous crop (e.g., leguminous crop), manure applications, irrigation water, and atmospheric deposition (Cassman et al., 2002; Meisinger et al., 2008; Iowa Soybean Association, 2008a).

Method/Placement

The goal of appropriate method and placement of fertilizer is to provide nutrients to plants for rapid uptake and to reduce the potential for losses to the environment. Studies have shown that NUE can be doubled under some conditions by placing fertilizers in the soil rather than "broadcasting" them on the surface (Malhi and Nyborg, 1991; Power et al., 2001). Liquid or gaseous forms of nitrogen can be injected directly into the soil with specialized equipment that is consistent with low-till systems. Solid forms can be broadcast on the surface and immediately incorporated into the soil with tillage equipment. Such placement reduces the risks of losses to the atmosphere and through surface runoff. The method of application can also reduce losses of nitrogen stemming from ammonia volatization (Meisinger and Randall, 1991; Peoples et al., 1995; Fox et al., 1996; Freney et al., 1981).

The impact of fertilizer placement on nitrous oxide emissions is less certain. Liu et al. (2006) found that injection of liquid urea ammonium nitrate at deeper levels resulted in 40-70 percent lower N_2O emissions than the rate associated with shallow injection or surface application. Some studies, however, have reported that incorporation into the soil increases N_2O emissions (Flessa and Beese, 2000; Wulf et al., 2002; Drury, 2006). Injection or incorporation could also increase nitrate leaching, especially where soils are coarse textured (Abt Associates, 2000).

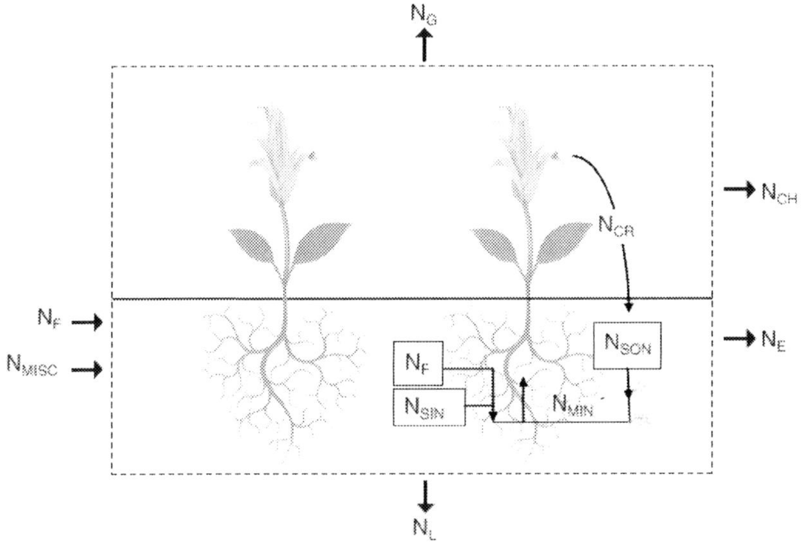

Nitrogen balance consists of N inputs of fertilizer and manure/legume N (N_F) and miscellaneous atmospheric deposition $_{(NMISC)}$; outputs of crop harvested N $_{(NCH)}$, N leaching (N_L), erosion (N_E), and gaseous losses (N_G); and internal N pools of crop residue N (N_{CR}), soil organic N $_{(NSON)}$, soil inorganic N $_{(NSIN)}$, and net N mineralization (N_{MIN}). Nitrogen use efficiency is the proportion of all N inputs (N_F and $_{NMISC)}$ that are removed in harvested crop biomass (NCH), contained in recycled crop residues (N_{CR}), and incorporated into soil organic matter $_{(NSON)}$ and inorganic N $_{(NSIN)}$ pools. The remainder is what is lost to the atmosphere through gaseous emissions (N_G), leaching (N_L), and erosion (N_E). The goal of nitrogen management is to reduce these losses through reductions in fertilizer inputs and through soil, water, fertilizer, and crop management that affects the cycling of nitrogen in the soil.

Source: USDA, Economic Research Service using data from Meisinger et al., 2008.

Figure 2.1. Nitrogen balance and nitrogen use efficiency.

Timing

The research on improving NUE in crop production emphasizes the need for greater synchronization between crop nitrogen demand and the supply of nitrogen from all sources throughout the growing season (Doerge et al., 1991; Cassman et al., 2002; Meisinger and Delgado, 2002). Balancing supply and demand implies maintaining low levels of inorganic nitrogen in the soil when there is little plant growth and providing sufficient inorganic nitrogen fertilizer during periods of rapid plant growth (Doerge et al., 1991; Alva et al., 2005). For example, the corn plant's need for nitrogen is not very high until about 4 weeks after it emerges from the ground, which typically falls in June through July in the major corn-producing States (Baker, 2001). Ideally, to ensure that growing crops have adequate N and to minimize losses from the soil, a farmer could split nitrogen applications or "spoon feed" nitrogen when using center-pivot sprinkler irrigation systems from June through July-August, using information from soil tests and/or advanced remote sensing techniques (Bausch and Delgado, 2003). Though splitting nitrogen applications is seen as an effective way to increase NUE and reduce nitrogen losses to the environment, several factors must first be considered: workload, seasonal fertilizer price differences, the risk associated with not being able to apply at the right time, application costs, the possibility of compacting the soil, and possible damage to growing crops (Doerge et al. 1991; Westermann and Kleinkopf, 1985; Westermann et al., 1988; Alva et al., 2005; Delgado and Bausch, 2005).

Form

NUE is also influenced by the form of nitrogen fertilizer (Raun and Schepers, 2008; Freney et al., 1995). Some of the more widely used nitrogen fertilizer forms include anhydrous ammonia (gas), urea (solid), UAN (liquid), and manure (solid). These forms vary in how quickly they can be transformed into nitrate, which is what crops actually use. The closer in time the fertilizer is applied to when the crop needs it, the faster it needs to be transformed into nitrate. A mismatch of fertilizer form with appropriate timing can lead to large environmental losses and poor yields.

Manure Effects

Manure is an important source of N, but it poses challenging management problems (Eghball et al., 2002; Kirchmann and Bergstrom, 2001; Davis et al., 2002). The nitrogen content of manure depends on the animal type and the method of manure storage (Davis et al., 2002; Eghball et al., 2002), and nitrogen content may be inconsistent from batch to batch (Davis et al., 2002). Manure is more difficult to handle than inorganic nitrogen fertilizers, and, if in solid form, is difficult to apply uniformly. Most of the nitrogen content of manure is in the organic form and has to be mineralized before crops can use it. Since the transformation process depends on manure type, soil, and weather conditions, it is more difficult to control soil nitrate levels relative to crop needs when manure is applied than when other forms are applied (Eghball et al., 2002; Power et al., 2001). Consequently, controlling environmental losses from manured fields is more difficult than from fields using commercial fertilizer.

Off-Site Practices That Capture Nitrogen

Off-field conservation measures can be used in conjunction with onfield nitrogen management to either capture reactive nitrogen in biomass or convert it to inert N_2 through denitrification.

Examples of off-site practices include vegetative buffers or filters and restored and constructed wetlands (Hefting et al., 2003; Jacobs and Gilliam, 1985; Lowrance et al., 1984). Buffers and wetlands reduce nitrogen loads to water through plant uptake, microbial immobilization and denitrification, soil storage, and groundwater mixing (Pionke and Lowrance, 1991; Lowrance et al., 1997; Hey et al., 2005; Mayer et al., 2005).

Buffers can remove nitrogen from both surface flow and groundwater (Mayer et al., 2005; Angier et al., 2001; Randall et al., 2008; Mitsch and Day, 2006). The effectiveness of vegetative buffers depends on the size of the buffer, the density of vegetation, and hydrologic conditions within the buffer zone (Dosskey et al., 2005; 2007).

Based on a wide range of studies, Mayer et al. (2005) estimate that buffers can remove about 74 percent of the nitrogen passing through the buffer root zone. However, in many areas of the country where tile drains are used to control the water table, especially in the Corn Belt, subsurface flows pass below the root zone and are not filtered by vegetative buffers.

Restored wetlands have been shown to be effective at reducing the transfer of nitrogen from agricultural land to water bodies (Jansson et al., 1994) and have been proposed as a technique to remove reactive nitrogen from the environment (Hey et al., 2005; Mitsch and Day, 2006). Wetland vegetation uptakes nitrogen, and wet soils enhance denitrification.

The effectiveness of wetlands as a filter of reactive nitrogen depends on their size, seasonal weather conditions, and hydrologic characteristics. Wetlands also provide a host of other ecosystem services that are valued by society, such as wildlife habitat and carbon sequestration.

State of Nitrogen Management on Cropland

Nitrogen Management on U.S. Cropland

Data on the nutrient management practices of U.S. producers of barley, corn, cotton, oats, peanuts, sorghum, soybeans, and wheat (table 3.1) are derived from USDA's Agricultural Resource Management Survey (ARMS) (see box, "Agricultural Resource Management Survey"). The basic practices for improving NUE are agronomic application rate, appropriate timing of applications, and proper placement (USDA, NRCS, 2006). For the purposes of this analysis, these practices are defined as follows:

Table 3.1. Crops, ARMS Phase II reference years, States surveyed, commodities, and nitrogen uptake per unit of crop yield

Crop	Reference year	States surveyed	Commodity	Lbs N per unit	Unit
Barley	2003	CA, ID, MN, MT, ND, PA, SD, UT, WA, WI, WY	grain	0.9	bushel
Corn	2005	CO, GA, IL, IN, IA, KS, KY, MI, MN, MO, NE, NY, NC, ND, OH, PA, SD, TX, WI	grain silage	0.8 7.09	bushel ton
Cotton	2003	AL, AZ, AR, CA, GA, LA, MS, MO, NC, SC, TN, TX	lint plus seed	15.19	bale
Oats	2005	IL, IA, KS, MI, MN, NE, NY, ND, PA, SD, TX, WI	grain	0.59	bushel
Peanuts	2004	AL, FL, GA, NC, TX	nuts with pods	0.04	pound
Sorghum	2003	CO, KS, MO, NE, OK, SD, TX	grain	0.98 14.76	bushel ton
Soybeans	2006	AR, IL, IN, IA, KS, KY, LA, MI, MN, MS, MO, NE, NC, ND, OH, SD, TN, VA, WI	beans	3.55	bushel
Wheat Winter Other spring Durum	2004	CO, ID, IL, KS, MI, MN, MO, MT, NE, ND, OH, OK, OR, SD, TX, WA	grain grain grain	1.13 1.39 1.29	bushel bushel bushel

Notes: N = nitrogen. ARMS = USDA's Agricultural Resource Management Survey. The nitrogen uptake coefficients are from Lander et al. (1998). The coefficients for soft (1.02 lbs/bushel) and hard (1.23 lbs/bushel) winter wheat were averaged because the type of winter wheat produced was not available. To download estimates based on these data, or to learn more about the surveys, go to www.ers.usda.gov/data/arms/beta.htm.

Source: USDA, Economic Research Service using data from USDA's Agricultural Resource Management Survey (2003-06) and Lander et al. (1998).

Agricultural Resource Management Survey

USDA's Agricultural Resource Management Survey (ARMS) is an annual survey of farm and ranch operators administered by ERS and the National Agricultural Statistics Service (NASS). ARMS gathers data on field-level production practices, farm business accounts, and farm households. ARMS is a multiple-phase survey. In the fall, NASS interviews producers of major commodities, such as feed grains, food grains, or cotton, to collect information about production practices and land use on select fields. In the spring, NASS re-interviews farmers who successfully completed the fall survey. Spring data collection focuses on the structural and economic characteristics of the farm business and farm operator household. This approach helps link commodity production activities and conservation practices with the farm business and operator household.

Each phase of ARMS contains multiple versions of the survey questionnaire. The commonality of questions across versions provides one facet of data integration. In the fall data collection, the target commodity distinguishes questionnaires.

- ***Rate.*** Applying no more nitrogen (commercial and manure) than 40 percent more than that removed with the crop at harvest, based on the stated yield goal, including any carryover from the previous crop. This approach is consistent with a more traditional approach for estimating N rate recommendations (Millar et al., 2010) and is also the criterion used by NRCS in its assessment of conservation practices in the Upper Mississippi Basin (USDA, NRCS, 2010). Crop uptake coefficients are from NRCS (Lander et al., 1998, table 3.1). This agronomic rate accounts for unavoidable environmental losses that prevent some of the nitrogen that is applied from actually reaching crops.
- ***Timing.*** Not applying nitrogen in the fall for a crop planted in the spring.
- ***Method.*** Injecting (placing fertilizer directly into the soil) or incorporating (applying to the surface and then discing the fertilizer into the soil) nitrogen rather than broadcasting on the surface without incorporation.

Form also plays a role in nitrogen management for improving NUE. However, available data do not allow for an assessment of the form of nitrogen fertilizer applied.

In this report, we evaluate nitrogen management only during the survey year covered by ARMS data. The loss of nitrogen to the environment in a particular year is mostly a function of current and not past management decisions. However, current management decisions have to account for past management, such as planting of a legume. The amount of commercial nitrogen applied is readily available from the ARMS responses; however, the amount of manure nitrogen must be estimated. We base these estimates on the quantity of raw manure applied, the form of the manure (liquid or solid), and the animal source of the manure. We also note whether the previous crop was a legume so as to account for the potential carryover of nitrogen.

A farm can fall into one of eight nitrogen management categories, defined by the three management decisions in a particular year:

1. All of the criteria are followed.
2. The rate and timing criteria are followed.
3. The rate and method criteria are followed.
4. The timing and method criteria are followed.
5. Only the rate criterion is followed.
6. Only the timing criterion is followed.
7. Only the method criterion is followed.
8. None of the criteria are followed.

How Many Acres Treated with Nitrogen Met the Criteria for Best Management Practices?

Because the crops covered in the analysis were surveyed in different years, we specify a reference year, 2006, to examine the extent to which best nitrogen management practices are being followed. Weights are calibrated so that the weighted sums of acres planted by the surveyed crop producers match USDA's published estimates of planted acres for 2006 (USDA, NASS, 2008). This provides reasonable baseline estimates under the assumption that the percentages of planted and treated acres and application rates by management category were stable between the survey reference years (see table 3.1) and 2006. We maintain this assumption throughout the analysis.

Sixty-nine percent of the 242 million acres planted to barley, corn, cotton, oats, peanuts, sorghum, soybeans, and wheat in 2006 were estimated to be treated with commercial and/or manure nitrogen (table 3.2). Corn accounted for an estimated 45 percent of the 167 million crop acres treated with nitrogen and 65 percent of the 8.7 million tons of nitrogen applied to these crops during 2006.

The application rate criterion was not met on over 53 million acres treated with nitrogen (32 percent). Cotton had the highest percentage of treated acres not meeting the rate criterion (47 percent), followed by corn (35 percent). However, corn accounted for 50 percent of all treated crop acres not meeting the rate criterion.

The timing criterion was not met on over 40 million treated acres (24 percent). About 34 percent of treated corn acres received commercial and/or manure nitrogen in the fall. These corn acres account for over 64 percent of all treated crop acres not meeting the timing criterion.

Table 3.2. Planted and nitrogen-treated acres, nitrogen applied, and the shares of treated acres and applied nitrogen that did not meet the rate, timing, or method criteria, by crop, 2006

	Total			Did not meet rate		Did not meet timing		Did not meet method	
Crop	Planted acres	Treated acres	Tons N	Treated acres	Tons N	Treated acres	Tons N	Treated acres	Tons N
	Thousands			Percent					
Barley	3,452	3,176	98	14	23	20	20	25	25
Corn	78,327	76,052	5,799	35	46	34	26	37	20
Cotton	15,274	12,566	591	47	61	18	11	32	24
Oats	4,168	2,748	93	33	49	28	32	42	41
Peanuts	1,243	737	14	1	7	16	11	39	29
Sorghum	6,522	5,370	220	24	31	16	16	27	21
Soybeans	75,522	16,827	248	3	31	28	56	45	43
Wheat	57,344	49,808	1,766	34	50	11	12	37	32
Total	241,852	167,285	8,829	32	47	24	23	37	24

Notes: N = nitrogen. These estimates are based on weighted sums, where the weights were calibrated so that the sums of planted acres for each crop based on the survey data match published estimates of planted acres for 2006 (USDA, 2008).

Source: USDA, Economic Research Service using data from USDA's Agricultural Resource Management Survey (2003-06), Phase II. See table 3.1 for details.

Nitrogen was not incorporated/injected on over 61 million treated crop acres (37 percent). These acres received 24 percent of all applied nitrogen. Soybeans (45 percent) had the highest percentage of acres not meeting the method criterion. However, corn accounted for about 46 percent of all treated acres not meeting the method criterion.

Corn acres make up nearly half of all acres that are in need of some type of improvement in nitrogen management, in that at least one of the three criteria is not met (fig. 3.1). Any policy aimed at improving nitrogen use efficiency would have to consider the factors driving management decisions in corn production.

From a regional standpoint, the Corn Belt and Northern Plains dominate in terms of cropland not meeting the management criteria (figs. 3.2, 3.3). Not coincidentally, these are the primary corn-growing areas in the United States. However, in terms of nitrogen application in excess of the criterion rate, the Corn Belt and Lake States receive the greatest amounts of excess nitrogen (fig. 3.4).

As described in the previous chapter, NUE is highest when all three management criteria are met. Table 3.3 shows the percentage of treated acres in each nitrogen management category, as well as the degree to which excess nitrogen is applied in relation to the rate criterion. About 35 percent (58 million acres) of the treated acreage meet all three criteria. Corn has the smallest percentage of treated acres meeting all three criteria (30.4 percent). Because of the large amount of cropland planted to corn, this represents about half of all crop acres needing improvement in nitrogen management (rate, timing, or method). Only 4.2 percent of all treated acres do not meet any of the three criteria.

About 47 percent of all treated crop acres meet the method and timing criteria. Most of the acres exceeding the rate criterion do so by less than 50 percent. For example, about 14 percent of corn acres receive applications of 10 percent or less over the criterion rate. Reducing application rates on these acres so that the rate criterion is met would mean that nearly 80 percent of all corn acres would meet the rate criterion and that 35 percent of corn acres would meet all three criteria.

It should be noted that our findings differ somewhat from those reported by USDA's Conservation Effects Assessment Project (CEAP) assessment of the Upper Mississippi River Basin (USDA, 2010). The CEAP study reports smaller percentages of cropland meeting the nitrogen management criteria. The ERS study, however, examines nitrogen management for only the survey year. The CEAP analysis looks at nutrient management practices over an entire crop rotation, which may run from 2 to 5 years (see box, "CEAP Analysis of Nitrogen Management in the Upper Mississippi River Basin"). All three criteria had to be met in each year of the rotation for CEAP to consider the cropping system as having met the nitrogen management goal. The CEAP approach is stricter than that used by ERS.

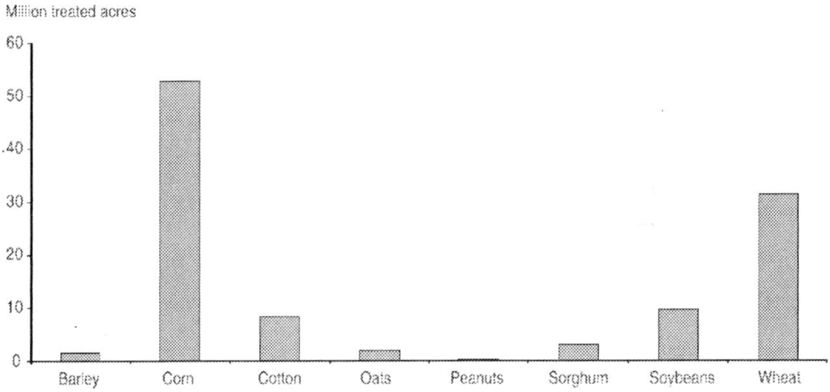

Source: USDA, Economic Research Service using data from USDA's Agricultural Resource Management Survey (2003-06), Phase II. See table 3.1 for details.

Figure 3.1. Acres treated with commercial and/or manure nitrogen not using nitrogen best management practices, 2006.

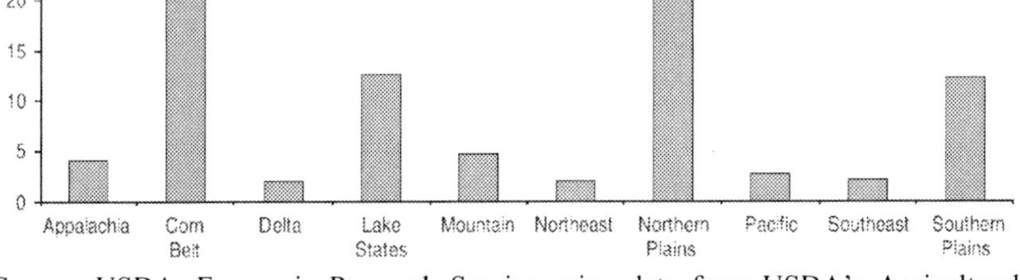

Source: USDA, Economic Research Service using data from USDA's Agricultural Resource Management Survey (2003-06), Phase II. See table 3.1 for details.

Figure 3.2. Acres treated with commercial and/or manure nitrogen not using nitrogen best management practices, by region, 2006.

Manure Use

Previous research has indicated that farms with animals tend to overapply nutrients to crops, primarily because of the large amount of manure produced on the farm needing disposal (Ribaudo et al., 2003; Gollehon et al., 2001). ARMS data provide additional evidence that manure use is associated with overapplication of nutrients. About 10 percent of crop acres treated with nitrogen (treated acres) received manure. Ninety-three percent of treated acres receiving manure did not meet all three criteria, compared with 62 percent of treated acres not receiving manure (table 3.4). Most of the cropland receiving manure was used to grow corn (72 percent). Over 95 percent of the corn acres receiving manure did not meet all three criteria, compared with 65 percent for corn acres not receiving manure.

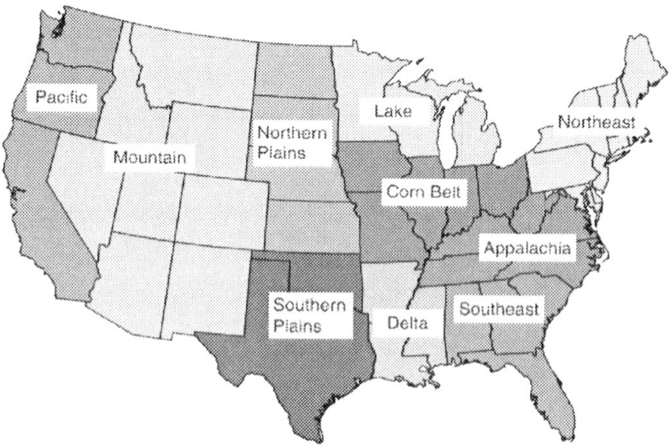

Source: USDA, Economic Research Service.

Figure 3.3. USDA farm production regions.

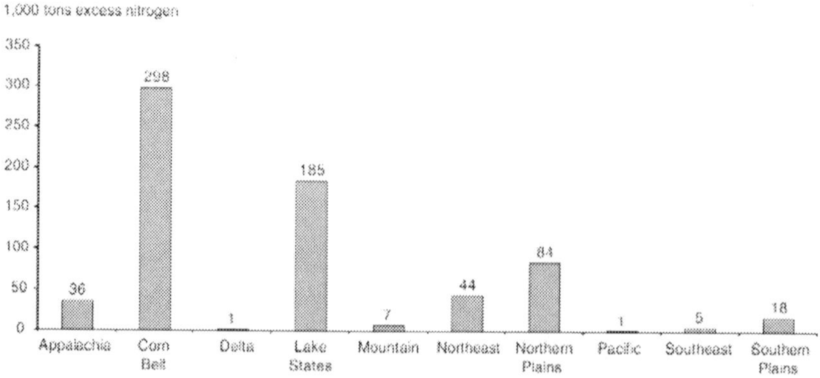

Note: Criterion rate defined as nitrogen removed at harvest plus 40 percent.

Source: USDA, Economic Research Service using data from USDA's Agricultural Resource Management Survey (2003-06), Phase II. See table 3.1 for details.

Figure 3.4. Total nitrogen applications above criterion rate by region, 2006.

Other Considerations

The environmental impacts of low nitrogen use efficiency on the environment can be affected by different land management practices, such as the presence of underground tile drains and the use of filter strips or riparian buffers. Tile drainage plays a role in nitrogen losses from fields (David et al., 2010). Tile drainage lowers the water table, enabling fields that would otherwise be wet part of the year to be intensively cropped. These drained soils tend to be highly productive. Tiles, however, provide a rapid conduit for soluble nitrate, effectively bypassing any attenuation that may occur in the soil. ARMS data indicate that nearly 26 percent of treated cropland is tiled, most of this in corn production (table 3.5). Of particular interest is the degree to which nitrogen management on this vulnerable cropland is not using nitrogen BMP. ARMS data indicate that about 71 percent of tiled acres do not meet all three nitrogen management criteria. Most of these acres are in corn production. Much of the tile-drained cropland is located in the Mississippi River Basin, which has implications for hypoxia in the Gulf of Mexico.

Land management practices can mitigate nitrogen losses from fields. The use of filter strips or riparian buffers can reduce the amount of nitrogen lost to surface water bodies. Less than 10 percent of treated crop acres not meeting the rate, timing, or method criteria have filter strips that could reduce losses in runoff and subsurface flows (table 3.6). For corn, about 11 percent of acres not using nitrogen BMPs have filter strips that could mitigate losses to water, but significant improvements could still be made. Filter strips, however, do not address atmospheric losses and may not be effective if not sited or managed appropriately. In addition, buffers would be ineffective on the 26 percent of treated cropland that is tile drained.

Nitrogen Management on U.S. Corn

A high percentage of crop acres meet at least some of the nitrogen management criteria (see table 3.3). Corn, however, meets all three criteria least often. Corn is the most intensive user of nitrogen and the most widely planted crop. Improvements in rate, timing, and/or application method are needed on 70 percent of corn acres to improve NUE. In addition, growth in corn demand due to the biofuels mandate suggests that corn acreage may increase in the future, along with the intensity of corn production. Together, these factors could increase reactive nitrogen emissions to the environment unless nitrogen use efficiency is improved.

Table 3.3. Percent treated acres by management category, crop, and degree of excess application, 2006

	Timing or method criteria met			
Rate criterion status	Timing and method	Timing	Method	Neither
	Percent of treated acres			
At or less than criterion rate				
Barley	52.0	16.4	13.0	4.3
Corn	30.4	15.0	12.0	6.2
Cotton	32.9	11.6	6.5	2.3
Oats	33.8	13.5	8.0	11.0
Peanuts	53.5	29.7	7.0	8.7
Sorghum	44.5	18.4	9.5	3.3
Soybeans	43.0	27.7	9.8	16.1
Wheat	36.8	22.2	5.2	1.7
Total	34.8	18.3	9.2	5.5
0 -10% over rate				
Barley	1.8	1.1	0.7	0.1
Corn	4.6	2.0	4.2	3.3
Cotton	6.6	3.5	3.7	0.7
Oats	1.2	1.2	0.0	0.1
Peanuts	0.3	0.0	0.0	0.0
Sorghum	3.7	0.3	1.1	0.1
Soybeans	0.0	0.1	0.1	0.0
Wheat	4.4	2.5	0.8	0.1
Total	4.1	2.0	2.5	1.6
10-50% over rate				
Barley	3.9	1.7	1.0	0.9
Corn	4.6	5.0	2.3	2.8
Cotton	10.4	7.2	1.8	0.8
Oats	6.4	2.3	0.5	1.6
Peanuts	0.1	0.0	0.0	0.0
Sorghum	6.6	1.9	1.7	0.3
Soybeans	0.0	0.2	1.1	0.6
Wheat	8.9	5.3	2.6	0.1
Total	5.9	4.5	2.1	1.5
50-100% over rate				
Barley	1.1	0.3	0.1	0.2
Corn	0.7	1.0	0.4	1.1
Cotton	3.3	4.0	1.3	0.5
Oats	3.6	1.9	0.4	1.3
Peanuts	0.0	0.4	0.0	0.0
Sorghum	1.7	0.3	0.0	0.0
Soybeans	0.1	0.1	0.1	0.2
Wheat	3.9	3.1	0.1	0.0
Total	1.9	1.7	0.3	0.6
Greater than 100% over rate				
Barley	0.2	0.0	0.1	0.2

	Timing or method criteria met			
Rate criterion status	Timing and method	Timing	Method	Neither
	Percent of treated acres			
Corn	0.6	0.3	1.2	0.8
Cotton	1.5	0.9	0.2	0.2
Oats	2.2	4.7	1.3	4.3
Peanuts	0.3	0.0	0.0	0.0
Sorghum	4.3	2.0	0.1	0.0
Soybeans	0.4	0.0	0.0	0.0
Wheat	0.2	1.7	0.1	0.1
Total	0.7	0.9	0.6	0.5
Total not meeting rate criterion	12.6	9.1	5.5	4.2

Notes: Figures in bold meet the rate criterion. See the notes to table 3.2.

Source: USDA, Economic Research Service using data from USDA's Agricultural Resource Management Survey (2003-06), Phase II. See table 3.1 for details.

CEAP Analysis of Nitrogen Management in the Upper Mississippi River Basin

Our assessment of nitrogen management on cropland using data from USDA's Agricultural Resource Management Survey (ARMS) has some similarities with the assessment of nutrient management on cropland in the Upper Mississippi River Basin (UMRB) conducted by the Conservation Effects Assessment Project (CEAP). The two studies also have some important differences. CEAP was initiated by USDA's Natural Resources Conservation Service, Agricultural Research Service, and Cooperative State Research, Education, and Extension Service (recently renamed the National Institute of Food and Agriculture). The goal of CEAP is to estimate conservation benefits from conservation investments and to provide research and an assessment on how to best use conservation practices in managing agricultural landscapes to protect and enhance environmental quality. The assessment of cultivated cropland in the UMRB is the first of a series of studies that will cover major crop-producing areas of the United States. Findings from the UMRB study are available at www.nrcs.usda. gov/technical/nri/ceap/umrb/index.html.

Both analyses assess baseline nitrogen management on cropland according to three criteria: rate, timing, and method. The definitions we used for each are based on those used in the CEAP analysis. Both studies used a survey to collect data on nutrient management practices. The major difference between our analysis using ARMS data and the CEAP analysis is how the criteria were applied.

ARMS collects information about cropping practices during a single crop year. Our analysis, therefore, based the assessment of nitrogen management on practices used to produce the crop sampled by the survey. The CEAP analysis focused on cropping systems, which could be up to 5 years in length and contain several different crops. Data were collected on production practices used each year of the crop rotation. CEAP used these data to evaluate the entire rotation, not just the crop grown during the year the survey was conducted. If the rate, timing, or method criteria were not met during any year of the rotation, then that sample point was identified as not meeting the nitrogen management criteria.

This approach is more stringent than the one used in our analysis. For example, assume corn and soybeans were on a 2-year rotation and that corn was grown during the year the ARMS and CEAP surveys were conducted. In our analysis, if the nitrogen application rate on corn met the rate criterion, then that corn sample was identified as such. In the CEAP study, the nitrogen application rate on both the corn and the previous year's soybean crops were assessed. If the application rate on corn met the rate criterion but excess nitrogen was applied to soybeans, then the rotation was identified as not meeting the criterion.

This leads to the CEAP assessment reporting a smaller percentage of crop acres meeting the rate criterion than we would report. Overall, the CEAP analysis reports fewer crop acres meeting the rate, timing, and method criteria than does the ERS report.

An examination of an additional year of survey data collected during the 2001 growing season and disaggregated regionally helps determine if management has undergone recent changes and if such changes vary by region. The share of corn acres not meeting the rate criterion declined from 41 to 35 percent between 2001 and 2005 (table 3.7).

This finding is in agreement with those of other reports on improving nitrogen use efficiency based on steady application rates and increased corn yields (Turner et al., 2007). Improvements in rate were seen in all regions except Appalachia and the Southeast. Notable improvements were seen in the Corn Belt, Lake States, and Northeast.

Timing and method, however, did not show similar improvements in the more recent data. For most regions, the percentage of corn acres not meeting these two criteria increased.

Table 3.4. Percent treated crop acres receiving commercial or manure nitrogen that did not meet the rate, timing, and method criteria, by crop, 2006

Crop	Planted acres	Treated acres	Acres treated with commercial N only		Acres treated with commercial and manure N		Acres treated with manure N only	
	Thousands		Percent of all treated acres	Percent vulnerable[1]	Percent of all treated acres	Percent vulnerable	Percent of all treated acres	Percent vulnerable
Barley	3,452	3,176	94	45	4	96	2	89
Corn	78,327	76,052	84	65	14	96	2	91
Cotton	15,274	12,566	96	67	3	85	1	29
Oats	4,168	2,748	78	59	9	88	13	92
Peanuts	1,243	737	93	46	5	52	2	41
Sorghum	6,522	5,370	98	55	1	98	1	49
Soybeans	75,522	16,827	85	51	2	100	13	91
Wheat	57,344	49,808	99	63	1	92	0	28
Total	241,852	167,285	90	62	7	96	3	86

[1]Vulnerable acres are those not meeting the rate, timing, and method criteria.

Notes: N = nitrogen. See notes to table 3.2. These estimates were weighted by the total amount of nitrogen applied by management category.

Source: USDA, Economic Research Service using data from USDA's Agricultural Resource Management Survey (2003-06), Phase II. See table 3.1 for details.

Table 3.5. Nitrogen-treated acres with tile drainage that did not meet the rate, timing, or method criteria by crop, 2006

	Treated acres		
Crop	Total	With tile drains	Tile-drained acres that do not meet the rate, timing, or method criteria
	Thousands		Percent
Barley	3,176	42	43
Corn	76,052	34,738	70
Cotton	12,566	583	71
Oats	2,748	216	66
Peanuts	737	40	44
Sorghum	5,370	46	43
Soybeans	16,827	5,690	69
Wheat	49,808	1,644	94
Total	167,285	43,000	71

Notes: See notes to table 3.2.

Source: USDA, Economic Research Service using data from USDA's Agricultural Resource Management Survey (2003-06), Phase II. See table 3.1 for details.

Table 3.6. Nitrogen-treated acres not meeting the rate, timing, or method criteria that have filter strips, by crop, 2006

Crop	Number of acres not meeting rate, timing or method criteria	No. of acres not meeting rate, timing, or method criteria with filter strips	% of acres with filter strips not meeting criteria
Barley	1,523	68	4
Corn	52,910	5,909	11
Cotton	8,432	397	5
Oats	1,818	99	5
Peanuts	343	42	12
Sorghum	2,983	64	2
Soybean	9,600	475	5
Wheat	31,475	2,530	8
Total	109,084	9,584	9

Notes: See notes to table 3.2.

Source: USDA, Economic Research Service using data from USDA's Agricultural Resource Management Survey (2003-06), Phase II. See table 3.1 for details.

Changing Management May Result in Environmental Tradeoffs

Changing management practices may improve nitrogen use efficiency, but the environmental outcomes may not always be desirable. We use the new Nitrogen Loss and Environmental Assessment Package with GIS (Geographic Information System) capabilities (NLEAP-GIS) model to assess how changes in nitrogen management practices on corn affect the losses of nitrate (to water), nitrous oxide (to air), and ammonia (to air) (Shaffer et al., 2010; Delgado et al., 2010a). Of particular interest is the extent to which tradeoffs in environmental outcomes might occur as overall nitrogen use efficiency is improved. See appendix 2 for more details on NLEAP.

Because NLEAP is a field-level model, we selected eight different soils in four States (Arkansas, Ohio, Pennsylvania, and Virginia) to assess changes in nitrogen emissions to the environment from management changes in nonirrigated corn production.[1] Four of the soils are type A or B soils (well drained), and four are type D soils (relatively poorly drained). For each soil, we examined two rotations (corn-corn and corn-soybeans), two tillage practices (conventional and no-till), and two sources of nitrogen (inorganic fertilizer and inorganic fertilizer + animal manure). The slopes for these soils were 0 to 6 percent, with low erosion potential.

For each soil/rotation/tillage/nitrogen-source combination, eight different scenarios were modeled with NLEAP, each representing one of the combinations of nitrogen management criteria outlined in table 3.4. Therefore, 64 different scenarios were modeled for each soil.

Table 3.7. Nitrogen-treated acres and the shares that did not meet the rate, timing, or method criteria for corn, 2001 and 2005

Region	Treated acres		Did not meet rate		Did not meet timing		Did not meet method	
	2001	2005	2001	2005	2001	2005	2001	2005
	Thousands		Percent of treated acres					
Appalachia	1,925	2,118	52	66	12	16	56	78
Corn Belt	35,087	39,145	46	38	41	41	39	34
Lake States	12,965	13,958	46	34	37	41	36	30
Mountain	1,243	1,018	18	14	9	20	33	50
Northeast	2,696	2,477	42	32	39	40	54	53
Northern Plains	16,962	18,293	27	28	10	15	36	45
Southeast	280	286	39	50	27	29	41	55
Southern Plains	1,708	2,109	31	32	45	38	33	18
Total	72,868	79,404	41	35	32	34	38	37

Notes: In both years, corn producers were surveyed in Colorado (Mountain); Kansas, Nebraska, North Dakota, and South Dakota (Northern Plains); Texas (Southern Plains); Michigan, Minnesota, and Wisconsin (Lake States); Illinois, Indiana, Iowa, Missouri, and Ohio (Corn Belt); New York and Pennsylvania (Northeast); Kentucky and North Carolina (Appalachia); and Georgia (Southeast). These estimates are based on weighted sums, with the weights recalibrated so that the weighted sums of planted acres for each crop based on the survey data match estimates for 2001 and 2005 (USDA, 2008).

Source: USDA, Economic Research Service using data from USDA's 2001 and 2005 Agricultural Resource Management Survey, Phase II.

A recommended application rate was specified for each soil/cropping system combination, based on local agronomic recommendations, as described by Espinoza and Ross (2008) for Arkansas, Alley et al. (2009) for Virginia, Beegle and Durst (2003) for Pennsylvania, and Vitosh et al. (1995) for Ohio. For the purposes of this analysis, overapplication was set at 75 percent more than the recommended rate (at the upper end of overapplication found in the ARMS data and reported in table 3.3). For example, if the recommended rate was 132 pounds of N per acre, the overapplication scenario used 231 pounds (see app. 2).

The modeled policy goal is that all three nitrogen management criteria be met. For demonstration purposes, we used the NLEAP results to assess the potential emissions tradeoffs when method, timing, timing and method, or rate BMPs are adopted by corn farmers. For example, to evaluate the tradeoff when timing is improved (rate and method criteria are already met), we compare the NLEAP results for the rate and method BMPs with the results for the rate, timing, and method BMPs. Each cropping system is evaluated separately. Because of the volume of results for the eight soils modeled, we present only those from the two soils in Ohio (tables 3.8a-d). Results for the other States.are similar, in terms of the direction of changes.

Table 3.8a. Changes in nitrogen losses resulting from improvements in nitrogen management, NLEAP estimates -Ohio - Type A soil - conventional tillage

Management improvement	Without manure Criterion rate=132 pounds N per acre				With manure Criterion rate=198 pounds N per acre*			
	Total	NO_3 [5]	N_2O [6]	NH_3 [6]	Total	NO_3	N_2O	NH_3
	Pounds of N per acre							
Continuous corn								
Method[1]	-32.8	7.0	-1.7	-38.1	-17.0	24.6	-1.2	-40.4
Timing[2]	-16.6	-17.4	0.8	+	-16.6	-17.6	1.0	+
Method+ timing[3]	-33.0	-9.1	0.4	-23.7	-18.6	11.4	0.8	-30.8
Rate4	-69.3	-50.6	-0.9	-17.7	-105.1	-81.0	-1.3	-22.9
	Criterion rate=102 pounds N per acre				*Criterion rate=153 pounds per acre**			
Corn-soybean								
Method[1]	-16.6	0.4	-0.8	-16.2	-14.7	3.8	-0.4	-18.1
Timing[2]	-5.7	-6.0	0.3	+	-5.2	-5.6	0.4	+
Method+ timing[3]	-13.1	-4.2	0.1	-9.0	-13.8	0.5	0.3	-14.6
Rate[4]	-15.7	-8.6	-0.4	-6.8	-37.2	-26.0	-0.6	-10.6

All the scenarios show the expected changes in total nitrogen losses, with reductions indicating improvements in NUE. The NLEAP results were consistent with the expectation that nitrogen emissions are minimized when all three criteria are met. Since nitrogen cycles through different forms and ecosystems, the long-term environmental benefits of reducing total nitrogen are clear. However, some of the tradeoffs between different forms of nitrogen could pose environmental problems. Adopting injection/incorporation always increases nitrate leaching, sometimes substantially (more than doubling leaching in some cases). Similarly, shifting applications from fall to spring (without changing application rate) reduces nitrate losses and total nitrogen losses but increases N_2O emissions as applications are shifted to generally warmer, wetter conditions, which is consistent with the findings of Delgado et al. (1996), Rochette et al. (2004), and Hernandez-Ramirez et al. (2009). Because of concerns over greenhouse gas (GHG) emissions, this outcome would have to be carefully considered when making recommendations to improve nitrogen use efficiency.

Adopting both method and timing again produces mixed results. NH_3 emissions are always reduced. Leaching is generally reduced, but in some cases where manure is used, it may increase. N_2O emissions almost always increase, from 5 to 50 percent, depending on the situation. In agreement with basic principles of nitrogen management, only reducing the application rate guarantees that losses of all three forms of reactive nitrogen are reduced (Mosier et al., 2002; Meisinger and Delgado, 2002). Based on these findings, a recommendation could be that in areas where leaching to drinking water sources is a concern, improvements in nitrogen use efficiency could focus on application rate reductions or improvements in timing.

Table 3.8b. Changes in nitrogen losses resulting from improvements in nitrogen management, NLEAP estimates – Ohio – Type A soil - no-till

Management improvement	Without manure Criterion rate=116 pounds N per acre				With manure Criterion rate=174 pounds N per acre*			
	Total	NO_3 [5]	N_2O [6]	NH_3 [6]	Total	NO_3	N_2O	NH_3
	Pounds of N per acre							
Continuous corn								
Method[1]	-29.6	5.6	-1.1	-34.1	-15.6	23.5	-0.3	-38.8
Timing[2]	-27.5	-28.6	1.1	+	-16.2	-17.3	1.1	+
Method+ timing[3]	-40.6	-20.8	1.1	-20.9	-27.3	0.2	1.3	-28.8
Rate[4]	-53.7	-37.3	-0.6	-15.8	-85.0	-63.8	-0.8	-20.3
	Criterion rate=86 pounds N per acre				Criterion rate=129 pounds N per acre*			
Corn-soybean								
Method[1]	-14.0	0.7	-0.8	-13.9	-12.7	4.7	-0.1	-17.3
Timing[2]	-9.9	-10.3	0.4	+	-8.6	-9.0	0.4	+
Method+ timing[3]	-14.9	-7.6	0.3	-7.6	-15.1	-2.2	0.5	-13.4
Rate[4]	-15.5	-9.5	-0.3	-5.7	-28.2	-18.7	-0.4	-9.1

Note:*Manure is applied every other year. Criterion rate is met on average over 2-year period. + indicates a positive but very small change. N = nitrogen. NO_3 = nitrogen trioxide. N_2O = nitrous oxide. NH_3 = ammonia.

[1] Timing and rate best management practices (BMP) are already in place.

[2] Method and rate BMPs are already in place.

[3] Rate BMP is already in place.

[4] No BMPs are in place.

[5] Nitrate loss to water (primarily through leaching but often ends up in surface water).

[6] Ammonia and nitrous oxide loss to atmosphere.

Source: USDA, Economic Research Service.

Summary

The survey data indicate that in 2006, all of the nitrogen management criteria were met on an estimated 35 percent of the crop acres treated with commercial and/or manure nitrogen.[2] In addition, a high percentage of treated acres met at least some of the nitrogen management criteria. Among all crops, corn met the criteria the least, and corn accounts for 50 percent of the treated acres upon which one or more improvements to management could be made to improve nitrogen use efficiency. Improvements in rate, timing, and/or method might be needed on 67 percent of corn acres.

Table 3.8c. Changes in reactive nitrogen losses resulting from improvements in nitrogen management, NLEAP estimates – Ohio - Type D soil - conventional till

	Without manure Criterion rate=132 pounds N per acre				With manure Criterion rate=198 pounds N per acre*			
Management improvement	*Total*	NO_3[5]	N_2O[6]	NH_3[6]	*Total*	NO_3	N_2O	NH_3
	Pounds of N per acre							
Continuous corn								
Method	-28.3	0.7	-5.0	-24.0	-20.0	12.9	-3.1	-29.8
Timing	-8.1	-9.4	1.3	+	-12.1	-13.5	1.4	+
Method+timing	-20.2	-7.6	1.2	-13.8	-17.2	4.6	1.7	-23.5
Rate	-56.3	-44.1	-1.8	-10.4	-91.3	-70.9	-3.0	-17.4
	Criterion rate=102 pounds N per acre				*Criterion rate=153 pounds N per acre**			
Corn-soybean								
Method	-14.7	0	-4.1	-10.6	-16.2	1.3	-2.2	-15.3
Timing	-1.9	-2.5	0.6	+	-2.7	-3.3	0.6	+
Method+timing	-6.8	-2.1	0.5	-5.2	-12.5	-0.4	0.8	-12.9
Rate	-9.3	-4.7	-0.7	-3.9	-27.8	-17.1	-1.4	-9.3

Note:*Manure is applied every other year. Criterion rate is met on average over 2-year period. + indicates a positive but very small change. N = nitrogen. NO_3 = nitrogen trioxide. N_2O = nitrous oxide. NH_3 = ammonia.

[1] Timing and rate best management practices (BMP) are already in place.

[2] Method and rate BMPs are already in place.

[3] Rate BMP is already in place.

[4] No BMPs are in place.

[5] Nitrate loss to water (primarily through leaching but often ends up in surface water).

[6] Ammonia and nitrous oxide loss to atmosphere.

Source: USDA, Economic Research Service.

NLEAP-GIS simulation results reported in the literature show that changing timing or method of application could potentially increase the loss of one type of nitrogen compound, even if total nitrogen emissions decline and NUE increases. NLEAP modeling indicates that only reducing application rates ensures that all nitrogen emissions decrease, in agreement with established principles of nitrogen management.

Table 3.8d. Changes in reactive nitrogen losses resulting from improvements in nitrogen management, NLEAP estimates –Ohio - Type D soil - no-till

Management improvement	Without manure Criterion rate=116 pounds N per acre				With manure Criterion rate=174 pounds N per acre*			
	Total	*NO_3*[5]	*N_2O*[6]	*NH_3*[6]	*Total*	*NO_3*	*N_2O*	*NH_3*
	Pounds of N per acre							
Continuous corn								
Method	-35.4	0.7	-1.4	-34.4	-25.8	13.6	-0.3	-39.1
Timing	-21.4	-22.0	0.6	+	-11.1	-11.8	0.7	+
Method+timing	-38.8	-18.3	0.6	-21.1	-32.2	-4.1	1.2	-29.3
Rate	-37.3	-20.4	-1.0	-15.9	-66.3	-44.2	-1.8	-10.4
	Criterion rate=86 pounds N per acre				*Criterion rate=129 pounds N per acre**			
Corn-soybean								
Method	-14.5	0.3	-0.8	-14.0	-16.0	1.6	0	-17.6
Timing	-7.2	-7.4	0.2	+	-6.2	-6.5	0.3	+
Method+timing	-13.3	-5.9	0.2	-7.6	-16.7	-3.7	0.6	-13.6
Rate	-10.1	-4.0	-0.4	-5.7	-20.4	-10.5	-0.7	-9.2

Note:*Manure is applied every other year. Criterion rate is met on average over 2-year period. + indicates a positive but very small change. N = nitrogen. NO_3 = nitrogen trioxide. N_2O = nitrous oxide. NH_3 = ammonia.

[1] Timing and rate best management practices (BMPs) are already in place.

[2] Method and rate BMPs are already in place.

[3] Rate BMP is already in place.

[4] No BMPs are in place.

[5] Nitrate loss to water (primarily through leaching but often ends up in surface water).

[6] Ammonia and nitrous oxide loss to atmosphere.

Source: USDA, Economic Research Service.

Policy Instruments for Nitrogen Reduction

Based on ARMS data, 65 percent of surveyed cropland, or 109 million acres, is in need of improved nitrogen management. Given nitrogen's effects on the environment, improving nitrogen management on vulnerable lands is a policy goal, both nationally and regionally.

Farmers adjust the management of their crops for a variety of reasons. Economic factors, such as input or output price changes, may lead to more (or less) careful use of nitrogen inputs. Farmers may also have to consider various policy-based incentives for adopting practices that improve nitrogen management. Over the years, policy instruments have been employed to improve the management of agricultural inputs and resources. USDA conservation programs rely primarily on subsidies for management practices and education. USDA also employs compliance mechanisms to protect wetlands and highly erodible soils. The U.S. Environmental Protection Agency (EPA) is using regulations to address nutrient management on certain confined animal feeding operations. A few States have used nitrogen fertilizer taxes to raise revenue for nutrient management programs. Such policy approaches may have a role to play in increasing the number of crop acres that meet the three nitrogen management criteria described earlier.

Provide Information (Education)

A lack of knowledge about their performance may be preventing farmers from using the most efficient nutrient management practices. Education is used to provide producers with information on how to farm more efficiently. Its success depends on alternative practices being more profitable to farmers than current practices (Ribaudo and Horan, 1999). Two practices that can lead to more efficient fertilizer use are soil testing and tissue testing. These tests provide information that reduces some of the uncertainty surrounding nutrient availability and enables producers to apply fertilizer at rates more consistent with plant needs and high nitrogen use efficiency.

ERS research supports previous findings that nitrogen testing is having the desired effect on nitrogen application rates for certain nitrogen users. Data from the 2001 and 2005 ARMS indicate that about 21 percent of corn farmers used a soil or tissue test as a basis for their level of nitrogen application (table 4.1). Farmers who used commercial nitrogen followed the recommendations closely. In our sample, their mean application rate of nitrogen was 136 lbs per acre, and the mean recommended rate based on a nitrogen soil test was 137 lbs per acre (table 4.2). Compliance with the soil test, however, was much worse for farmers who used both manure and commercial fertilizer. In their case, the recommended nitrogen application rate was 123 pounds per acre. And while farmers applied only 85 pounds per acre of commercial fertilizer, total nitrogen application rates were 175 pounds per acre when manure was added. We compared nitrogen application rates of those farmers who use soil N and tissue tests with those who do not using regression analysis that accounts for a number of production, land, and operator characteristics (see app. 3). Findings show that soil nitrogen testing has a statistically significant impact on nitrogen application rates. In the case of farmers who use commercial nitrogen exclusively, those who tested the soil applied 73.9 pounds per acre less than those who did not, all else equal. Other studies have found soil tests to be of similar effectiveness (Wu and Babcock, 1998; Musser et al., 1995). An information-based approach can meet nitrogen efficiency goals only if the information provided leads to increased profits for farmers (Ribaudo and Horan, 1999). As long as there are expectations that more efficient nitrogen management leads to increased risk or higher costs, then nitrogen management goals are unlikely to be met with information alone. However, information has proven valuable in support of other policy goals. Education can reduce the cost of adopting nitrogen BMPs required by regulation or funded through financial incentives. For example, Bosch et al. (1995) found that education affected the outcomes associated with a regulation requiring nitrogen testing in Nebraska. Producers did not use the information provided by testing unless they received education assistance.

Financial Incentives

Financial assistance is an important tool used in many USDA conservation programs to promote the adoption of BMPs. Program effectiveness depends on how farmers respond to the incentive being offered. When a farmer accepts a payment in return for adopting a management practice, he or she is signaling that the payment at least represents the economic cost of implementing the practice, sometimes referred to as the willingness-to-accept. Generally, only the producer knows the true cost. This makes it difficult for program managers to find the minimum payment rate that entices enough producers into the program to achieve the particular environmental goal at least cost.

Table 4.1. Factors influencing farmers' nitrogen fertilizer application decision

Application used	2001	2005
	Percent of farmers	
Soil or tissue test	18.8	27.0*
Crop consultant recommendation	13.0	17.6*
Fertilizer dealer recommendation	28.7	41.2*
Extension service recommendation	3.2	4.6*
Cost of nitrogen and/or expected commodity price	11.4	17.3*
Routine practice	70.9	71.7*
	Number	
Observations	1,646	1,344

*Statistically different from 2001 at the 1-percent level, based on pairwise two-tailed delete-agroup Jackknife t-statistics (Dubman, 2000)

Source: USDA, Economic Research Service using data from USDA's 2001 and 2005 Agricultural Resource Management Survey, Phase II, Cost of Production Practices and Costs Report.

Table 4.2. Influence of soil/tissue nitrogen testing on fertilizer application rates for corn, with and without manure use, 2001 and 2005

For farmers using a soil test	Required nitrogen based on expected yield[1]	Soil test recommended nitrogen	Commercial nitrogen applied	Total nitrogen applied (commercial + manure)
	Pounds of nitrogen per acre			
Commercial nitrogen with manure Observations = 154	152	123	85†	175†
Commercial nitrogen without manure Observations = 645	165	137	136	136

[1]Based on nitrogen removed in expected harvest plus 40 percent to account for unavoidable nitrogen losses.

†Means are statistically different from the recommended nitrogen amount at the 1-percent level, based on pairwise two-tailed delete-a-group Jackknife t-statistics (Dubman, 2000).

Source: USDA, Economic Research Service using data from USDA's 2001 and 2005 Agricultural Resource Management Survey, Phase II, Cost of Production Practices and Costs Report.

USDA's NRCS supports management practices that specifically address fertilizer application rate, timing, or method in their standards. The

Environmental Quality Incentives Program (EQIP) is the largest USDA program that provides producers with technical and financial assistance for implementing and managing BMPs on working farmland. Management practices supported by EQIP that can influence nitrogen use efficiency include nutrient management and waste utilization (for manure). Implementing a nutrient management plan directly affects measures of stewardship. Nutrient management planning addresses the amount, source, placement, form, and timing of the application of plant nutrients and soil amendments (USDA, NRCS, 2006). Further, the practice requires the application rate be based on an assessment of plant-available nitrogen developed through Land Grant University soil and tissue tests or recognized industry practices. Waste utilization guidelines specify that rates of application must be compatible with the soil's ability to absorb and hold the waste, and methods of incorporation are prescribed for liquid manure forms to prevent nutrients from rising to the surface.

Data from EQIP contracts in force for year 2008 show that participating farmers accepted an average payment of $8.88 per acre for adopting nutrient management (table 4.3). A higher per acre payment induced farmers to adopt a waste utilization practice ($14.75). Relatively few corn farm operations have livestock or a direct source of manure (organic) fertilizer, and, as reported later, the practice can be more costly to farmers than using commercial (inorganic) fertilizer.

A focus on the Corn Belt reveals variation in the accepted payments for the two practices (table 4.3). The variation may stem from cost differences within the region that are driven by local conditions, which, in turn, influence the State-level payment rate for the practice. To examine how management practices can affect a farm's cost of operations, we estimate a cost function using a generalized linear regression model estimated with 2001 ARMS data (see app. 4).[3] Model results show that several conservation practices have little effect, on average, on the cost of operation relative to other methods of management. For example, the difference in operation costs for farms using nutrient management and for farms not using these practices is not statistically significant.

Based on results from our cost analysis, we also find that using manure as a nitrogen source costs roughly $26.84 more per acre than using only commercial fertilizer. However, we observe a national average per acre EQIP payment for the waste utilization of $14.75, and only two States in the Corn Belt (Illinois and Indiana) have payment levels that approach the estimated cost figure. The results suggest that the EQIP rate is insufficient to entice farmers who are not using manure to begin doing so in an environmentally sensitive manner. However, farms with livestock or poultry need to dispose of the waste. Therefore, rather than be a practice by choice, waste utilization may be a practice that complements the necessary disposal of manure, and a payment that covers increased production costs may not be a necessary condition for the willingness to adopt the practice.

Table 4.3. Per acre average EQIP payments for conservation practices, 2008

		Corn Belt				
Practice	All States	Illinois	Indiana	Iowa	Missouri	Ohio
	Dollars per acre					
Nutrient management[1]	8.88	9.75	7.47	6.12	13.90	10.91
Waste utilization[2]	14.75	25.95	25.84		10.90	5.83

[1] Nutrient management planning addresses the amount, source, placement, form, and timing of the application of plant nutrients and soil amendments.

[2] Waste utilization guidelines specify that rates of application must be compatible with the soil's ability to absorb and hold the animal waste, and methods of incorporation are prescribed for liquid manure forms to prevent nutrients from rising to the surface.

Notes: Blank cells indicate no contracts for such practice in that State.

Source: USDA, Economic Research Service using contract data from USDA's Environmental Quality Incentives Program, fiscal years 1997-2008, payments made in fiscal year 2008.

Not all farmers require a cost share to adopt conservation practices. Cooper and Keim (1996) use farmer surveys to conclude that 12 to 20 percent of farmers may be willing to adopt practices such as split fertilizer applications and nutrient testing without financial assistance but do not do so because they lack information or are uncertain about the practices' economic performance. However, they also find that the adoption rate would not increase beyond 30 percent unless subsidy rates were substantially increased. A farmer's perception of the effectiveness of a practice can also influence the decision to adopt. Evidence from Lichtenberg and Lessley (1992) suggests that farmers may need more than a cost share to overcome perceptions of conservation practices and the state of environmental quality off-site.

In some cases, farmers are willing to adopt conservation practices that reduce profits if they believe that others will benefit from the subsequent change in environmental quality (Bishop et al., 2010; Chouinard et al., 2008). For example, based on survey responses from the State of Washington, Chouinard et al. (2008) conclude that farmers would be willing to forgo up to $4.52 (median value estimate) in per acre annual profits to implement soil-conserving stewardship practices.

The scope of a program's coverage is an important consideration for policymakers and program managers evaluating the adequacy of the financial incentives offered to program participants. In 2008, the financial incentives from EQIP encouraged farmers to enroll 4 million acres in the program's nutrient management practice. However, because participation in the program is voluntary, it is not known if the cropland most in need of treatment was enrolled.

We can use the data from EQIP and table 3.3 to estimate the cost to improve nitrogen use efficiency on those acres needing additional treatment. About 35 percent of all crop acres meet all three criteria, which means that over 108 million acres of cropland are not using nitrogen BMPs. Applying the average payment rate for nutrient management ($8.88 per acre) to all acres needing improved management implies annual EQIP payments of $959 million. However, the findings from Cooper and Keim (1996) suggest that higher rates would be needed to entice a sizable percentage of farmers to voluntarily enroll in a program. Assuming a payment rate 50 percent higher results in program expenditures of $1.4 billion. This is roughly the current annual budget for EQIP.

Given the potential cost of treating the entire 108 million acres of cropland not using nitrogen BMPs, which groups might be most important to address first? We previously reported that manure users generally apply much more total nitrogen to the field than farmers who exclusively apply commercial nitrogen. Providing financial assistance for nutrient management on the 7.7 million acres that received manure and failed to meet the rate criterion would cost between $68.4 and $103 million per year.

Off-Site Filtering for Reducing Nitrogen Losses from Fields

Similar to its efforts aimed at improving nitrogen use efficiency on working lands, the government can provide financial incentives for installing management practices that capture nitrogen after it leaves a field, primarily nitrogen in water. This analysis estimates and evaluates the cost effectiveness of two such measures, wetlands restoration and vegetative filter strips (VFS), assuming that funding is targeted to areas where nitrogen removal is likely to be most effective.

The Costs of Nitrogen Capture by Restoring Wetlands

Our analysis of wetlands restoration focuses on the Glaciated Interior Plains (GIP), where models of wetlands nitrogen removal have been developed. The GIP includes major parts or all of Ohio, Minnesota, Wisconsin, Michigan, Iowa, Illinois, and Indiana—major corn-producing States. This area is also an important source of nitrogen that reaches the Gulf of Mexico and contributes to the hypoxic zone (Goolsby et al., 2001; Robertson et al., 2009). Wetlands in other parts of the United States can also reduce nitrogen loadings. But, because of regional differences in ecosystems, we do not extrapolate our findings to other areas.

Wetlands once made up a large portion of land on the GIP (fig. 4.1). Water tables were lowered to facilitate crop production by installing underground tile and surface drainage systems. Such drainage systems become conduits for the rapid movement of nitrate from fields to water resources.

The costs of creating wetlands vary widely as do nitrogen removal rates on wetlands. Costs are driven by the cost of the land and the cost of restoring wetland ecosystems. Nitrogen removal depends on the rate of nitrogen inflow, nitrogen concentration, seasonal variations in flow, wetland size, and other factors.

We use the USDA Wetland Reserve Program (WRP) contract data for the GIP to estimate multinomial land and restoration cost functions (see app. 5). With these functions, we generate county-level cost estimates throughout the GIP. The objectives of the WRP are to enhance, restore, and preserve wetlands. As of October 1, 2009, the WRP enrolled 2.18 million acres, with wetlands in every State. Along with the land and restoration cost variables, the WRP contract data contain information on the size and the county location of each contract. The land (wetland easement) cost variable represents the difference between the agricultural value of the land and the value of the land with a wetland easement. The easement requires that the landowner maintain the health of the ecosystem. Data for other variables in our analysis come from the NASS agricultural census. Across the counties within the GIP, wetland easement costs range from $1,490 to $3,030 per acre, as generated by our estimated land cost function. Expected wetland restoration costs range from $506 to $602 per acre. Annualizing over perpetuity with a discount rate of 5 percent, we estimate that the median annual expected cost of restoring and preserving wetlands is $153 per acre per year (table 4.4). Because marginal costs are less than average costs, one can expect average per acre costs to be lower for larger wetlands and potentially more cost effective as a nitrogen filter, all other things being equal.

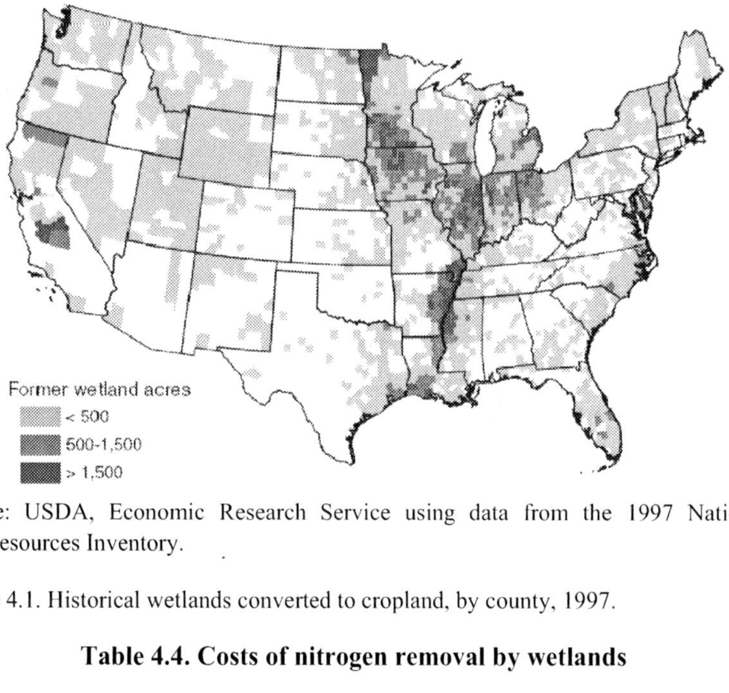

Source: USDA, Economic Research Service using data from the 1997 National Resources Inventory.

Figure 4.1. Historical wetlands converted to cropland, by county, 1997.

Table 4.4. Costs of nitrogen removal by wetlands

Wetland cost		N removal rate = 142 lbs/ac	N removal rate = 214 lbs/ac	N removal rate = 450 lbs/acre	N removal rate = 1,000 lbs/acre
	$/acre	*$/lb of N removed by wetland*			
Marginal cost	77	0.54	0.36	0.17	0.08
Average cost	153	1.08	0.71	0.34	0.15

Note: Because marginal costs are less than average costs, per acre costs would be lower for larger wetlands. N = nitrogen.

Source: USDA, Economic Research Service using data from Mitsch et al., 1999 (142 and 214 pounds per acre) and Crumpton et al., 2008 (450 and 1,000 pounds per acre).

Wetlands remove most nitrogen through denitrification (Crumpton et al., 2008), which converts nitrate to nitrous oxide (N_2O). However, there is a general belief, supported by a limited number of studies, that N_2O releases are a very small portion of nitrogen removal, even in wetlands with elevated nitrogen loadings (EPA, 2010b). Researchers estimate that N_2O accounts for between 0.13 and 0.30 percent of total annual wetland nitrogen loss (Hernandez and Mitsch, 2006; Crumpton et al., 2008). The reported rates of N_2O releases by wetlands are similar to estimated releases on cropland in the Midwest, so restoring wetlands is likely to have no net effect on N_2O emissions (Crumpton et al., 2008).

Crumpton et al. estimate that nitrogen loads to surface water could be reduced by 30 percent (~500 million pounds) in the Upper Mississippi and Ohio River basins with the addition of 0.5 to 1.1 million acres of strategically placed wetlands, for an average per acre reduction of 450 to 1,000 pounds per year. These removal rates assume an optimal placement of the restored wetlands—areas with a high water flow with high nitrogen concentrations. Mitch et al. (1999) estimate that wetlands in the Midwest remove 142 to 214 pounds per acre of nitrogen per year. The researchers assume that the wetlands are well constructed and placed, but their estimates are based on a wide range of nitrogen concentrations and hydrologic flows. Each study includes multiple wetlands and a variety of flow conditions and nitrogen concentrations.

The unit cost of nitrogen removal by wetlands, based on nitrogen removal rates of 450 to 1,000 pounds per acre per year reported by Crumpton et al. (2008), is \$0.08 to \$0.34 per pound (table 4.4). Based on the removal rates of 142 to 214 pounds per acre per year reported by Mitch et al. (1999), unit cost ranges from \$0.71 to \$1.08 per pound.

The Costs of Nitrogen Capture Using Vegetative Filter Strips

Vegetative filter strips present another off-field option for capturing and removing nitrogen from runoff and subsurface waters. The cost of a VFS tends to be lower than the cost of wetlands restoration. The VFS cost has two components: the opportunity cost of holding the land out of production and the cost of establishing cover (e.g., grasses, trees, or both). Cropland rental rates are an economic measure of the opportunity cost of taking cropland out of production. We assume that average cropland rental rates are equal to the economic return to land converted to a VFS. Based on the distribution of corn acreage reported in the 2005 ARMS and county-level rental data provided by NASS, the annual opportunity cost of converting corn cropland into a VFS is estimated at \$94 per acre.

We assume that the cost of establishing vegetative cover is about the same as establishing cover on land retired in USDA's Conservation Reserve Program (CRP). CRP data do not specify cost by cover type, but data do provide insights on the range of costs. Across the 25th, 50th, and 75th percentiles, cover costs are $16, $35, and $60 per acre. Because establishing forest cover is more costly, the lower percentile costs likely reflect the cost of establishing grasses. The cover cost is a one-time investment. We annualized this cost by assuming that it is to last for the foreseeable future and a 5-percent discount rate. Together, the land and cover cost would total approximately $95 to $97 per acre per year, with the higher estimate more likely representative of the use of forest cover. Mitch et al. (1999) tabulate several plot studies with a focus on the quantity of nitrogen removed across varying sizes of filter strips and levels of nitrogen inflow. They apply their findings to nitrogen runoff rates typical of those in corn-producing areas and estimate that properly designed forested riparian VFS will remove approximately 17.8 to 53.0 pounds of nitrogen per acre with strips ranging in width from 10 to 50 feet (Mitch et al., 1999, pg. 47). At an annual nitrogen runoff removal rate of 17.8 to 53.0 lbs per acre and a forested VFS cost of $97 per acre, VFS nitrogen removal costs are estimated to range from $1.83 to $5.45 per pound of nitrogen. The cost estimate is a weighted average across the corn-producing areas of the GIP. Results suggest that, within the GIP, wetlands can be much more cost effective at removing nitrogen than VFS, primarily because of their substantial nitrogen removal rates. Within corn-producing regions, especially in areas where fields are tile drained, water moves quickly through and passes under root zones, rendering VFS ineffective. On the other hand, VFS can be established in many landscape settings where wetlands cannot. The wide range in nitrogen removal rates by wetlands reflects, at least in part, the advantage of targeting wetlands to areas where they are likely to be more effective—areas where wetlands capture large quantities of water with high nitrogen concentration rates. But even the low nitrogen removal rates of 142 to 214 pounds per acre reported by Mitch et al. are three or more times the removal rates of VFS. Additionally, the rich wetland ecosystems have the potential of providing a greater array of environmental services than those delivered by VFS.

Participation in Emissions Trading Programs

An alternative to publicly provided financial incentives for adoption of conservation practices is for private markets to pay farmers to adopt management practices that produce ecosystem services valued by consumers (the public). Emissions trading uses markets to efficiently achieve pollution targets. The development of markets for ecosystem services is characterized by uncertainties about whether viable markets for public goods can exist, but the EPA and USDA are promoting emissions trading markets for water quality and greenhouse gases as a way of reducing the costs of meeting environmental goals. Agriculture has a potential role to play in both markets.

Water Quality Trading Program

The promise of emissions trading, along with the real-world success of air emissions trading, has led to the creation of water quality trading markets in a number of impaired watersheds. Under the Clean Water Act, point sources (e.g., factories, sewage treatment plants) were initially regulated through a nontradable permit system. A permit specifies how much of a particular pollutant the permit holder can discharge. Traditionally, permit holders were required to meet their permit obligations through their own effluent reductions. EPA policy guidelines on water quality trading now allow point sources to meet their Water Quality Based Effluent Limitation requirements through discharge reductions from other sources under certain conditions, including agricultural nonpoint sources (EPA, 2004). The guidelines encourage States to consider agriculture as a source of offsets in water quality trading programs, and a number of States are either implementing or considering water quality trading programs that allow point/nonpoint source trading. There appears to be many opportunities for point/nonpoint trading programs to be established. Almost 7,000 water bodies impaired by nutrients (pollutants produced by both point and nonpoint sources) have been listed under Section 303(d) of the Clean Water Act (EPA, 2009). To date, over 4,000 Total Maximum Daily Loads (TMDLs) have been developed to address 5,000 of these impaired waters. The presence of a TMDL is a basic requirement for a trading program, as it creates the demand for credits (Ribaudo et al., 2008). Agriculture is a major source of nutrients in most of the watersheds containing impaired waters (Ribaudo and Nickerson, 2009). The marginal cost of reducing nitrogen loss from cropland is generally less than the marginal cost of reducing nitrogen discharges from point sources (primarily sewage treatment plants) (Camacho, 1992; Shortle, 1990).

Forty water quality trading programs have been created in the United States since 1990 (Breetz et al., 2004). Fifteen include production agriculture as a potential source of credits for regulated point sources, most often for nutrients (nitrogen and phosphorus). However, point/nonpoint trading has not been very successful, at least in terms of the participation of potential traders and the number of trades between regulated sources and farms (Breetz et al.,

2004). Regulators designing point/nonpoint trading markets must contend with uncertainty about sources and levels of emissions, the effectiveness of best management practices, the water quality impacts of emissions from different sources, and farmer willingness to participate in a market driven by regulation (on point sources) (Hoag and Hughes-Popp, 1997; King, 2005; King and Kuch, 2003; Woodward and Kaiser, 2002; Ribaudo and Gottlieb, 2011; Horan and Shortle, 2011). The failure of current programs to perform as advertised can largely be attributed to failures of market design and program rules to adequately address these issues, or the high transactions from incorporating uncertainties into market design. One issue that has particular relevance for addressing nitrogen pollution is the baseline used for calculating credits. The EPA defines a baseline participation requirement as the pollutant control requirements that apply to a seller in the absence of trading (EPA, 2007). EPA suggests that practices generally accepted as good management define a baseline for agriculture, under an assumption that all farms would eventually adopt these practices voluntarily. Some practices that States have used in trading programs to define a baseline include the use of filter strips or a nutrient management plan (Wisconsin DNR, 2002; Pennsylvania DEP, 2008). However, the issue is that our survey data indicate that very few crop acres would meet these baseline requirements as the percentages of cropland with filter strips or nutrient management plans are only 6.8 and 5.0, respectively, meaning that most crop acres would not be able to participate in a trading program until the baseline requirements were met. If the incentives from a credit market are insufficient to induce farms that have not already voluntarily adopted the minimum set of practices to incur the cost of meeting the baseline requirement, then these farms will continue unabated discharge. This entry cost would therefore potentially limit participation and adversely affect the efficiency of the market (Ribaudo and Gottlieb, 2011; Ghosh et al., 2011).

Greenhouse Gas Mitigation

Another emissions market that might influence nitrogen management decisions in agriculture is an offset market for mitigating emissions of CO_2 and other greenhouse gases, such as nitrous oxide (N_2O). Nitrous oxide is a powerful greenhouse gas (310 times the global warming potential of CO_2 over 100 years) and can be emitted from fields receiving nitrogen fertilizer (see chapter 2). A trading program for nitrous oxide emissions would have many of the same design and implementation issues of point/nonpoint trading for water quality. One would expect that the use of models for predicting reductions, based on field and management characteristics, would figure heavily in any trading program. We use NLEAP results and ARMS cost data to determine changes farmers might make given the opportunity to participate in an offset market for N_2O reductions by producing credits and likely environmental tradeoffs. These analyses were conducted across different management scenarios and general hydrologic soils (e.g., well-drained soils with a large leaching potential versus poorly drained soils with a low leaching potential) from selected counties in Virginia, Ohio, Pennsylvania, and Arkansas.

For each soil, we identified the changes a farmer might make in nitrogen management practices to produce N_2O reductions (offset credits) at the lowest cost while meeting a requirement that total nitrogen emissions (the sum of NO_3, N_2O, and NH_3 losses) not increase. In other words, trading rules do not permit a management change that reduces N_2O but increases total nitrogen emissions. Changes in cost are defined as the difference in average variable costs (chemicals, fuel, and electricity) and value of lost production (changes in yields). We assumed farmers would maintain the same basic cropping system and alter timing, method, or application rate only. A description of NLEAP and the cost model and assumptions are presented in appendices 2 and 3. Table 4.5 summarizes the nitrogen management systems that farmers evaluated in the model would adopt to produce credits at the lowest cost, given baseline practices. For example, of the 64 farm types not meeting any of the criteria prior to a market ("None" in the baseline criteria column), 17 would reduce the application rate to the criterion rate, 10 would reduce the rate and inject/incorporate nitrogen, 1 would reduce the rate and apply nitrogen in the spring, and 36 would adopt all three management choices. The choice depends on the soil type, climate, rotation, tillage practice, and nitrogen source. The results highlight the importance of meeting the application rate criterion for reducing both N_2O and total reactive nitrogen. For all farms not meeting the rate criterion, reducing application rate either alone or in combination with another practices was selected to reduce N_2O. Method or timing was never the sole practice adopted by farms to reduce N_2O emissions. Model results also indicate that 148 of the 512 farming systems will not be able to reduce N_2O emissions by meeting the rate, timing, or method criteria. For example, none of the 64 farm types meeting the rate and method criteria at the start of a market can reduce N_2O emissions by also meeting the timing criterion. Table 4.6 provides more detail for one soil in Ohio. It shows the reduction in N_2O that would be generated for each decision a farmer in a particular baseline situation could make and credit revenue earned assuming a carbon price of $15 per ton of CO_2 equivalent.[4] The range of N_2O reductions presented here is similar to that found for the other soils modeled with NLEAP.

Table 4.5. Least-cost N management systems in corn production for reducing N_2O emissions for 512 model farms, assuming a credit price of $15 per ton of CO_2 equivalent, based on NLEAP modeling

Criteria[1] met after changing management	Method	Rate	Timing	Rate and method	Rate and timing	Timing and method	Rate, timing, and method	Total model farms
	Number of model farms							
Criteria[1] met in baseline								
None		17		10	1		36	64
Method		16		17	3		28	64
Rate		19		42			3	64
Timing					63		1	64
Rate and method				64				64
Rate and timing		3		23	1		37	64
Timing and method				31			33	64
Rate, timing, and method							64	64

[1]Criteria are appropriate rate, timing, and method of nitrogen application (see chapter 3).

Note: N = nitrogen. NLEAP = Nitrogen Leaching Environmental Analysis Project. N_2O = nitrous oxide. CO_2 = carbon dioxide. A total of 512 cropping systems are evaluated with NLEAP, 128 each in Arkansas, Ohio, Pennsylvania, and Virginia. Each defines a soil type (A or D), a rotation (continuous corn, corn soybeans), tillage practice (conventional, no-till), nutrient source (inorganic, manure+inorganic), timing of application (before planting, at/after planting), method (inject/incorporate, broadcast) and application rate (meet criterion, 75% over criterion).

Source: USDA, Economic Research Service.

Even though our sample of cropping conditions is very small, we believe we can still make some inferences from the results. We found that if the baseline system is not meeting the application rate criterion, application rate will be reduced to produce credits, either alone or in combination with timing or method; reducing the application rate is generally the most cost-effective means of reducing N_2O emissions. Adopting method and/ or timing BMPs alone cannot reduce N_2O emissions or can do so only by reducing overall nitrogen use efficiency, which is not permitted under our simulated market rules.

Table 4.6. How a corn farmer may change N management in a market for nitrous oxide (N_2O) greenhouse gas emissions with credit payments of $15/ ton of carbon dioxide equivalent, for a model Ohio farm on Ottoke soil

Baseline practice	Practices after N_2O credit offered	N_2O reduction	Credit revenue
		Pounds per acre	*Dollars per acre*
CC-CON-MF			
M	RTM	0.9	2.09
RM	No change	0.0	0.0
R	RM	0.3	0.70
RTM	No change	0.0	0.0
RT	RM	3.4	7.90
TM	RT	3.0	6.98
T	RT	4.4	10.23
NONE	RTM	0.8	1.86
CC-CON-OF			
M	RTM	0.3	0.70
RM	No change	0	0
R	RM	0.6	1.40
RTM	No change	0	0
RT	RTM	2.7	6.28
TM	RT	0.9	2.09
T	RT	3.1	7.21
NONE	RTM	0.8	1.86
CC-NT-MF			
M	RTM	0.2	0.46
RM	No change	0	0
R	No change	0	0
RTM	No change	0	0
RT	RTM	0.5	1.16
TM	RT	3.3	7.67
T	RT	2.8	6.51
NONE	RM	0.9	2.09
CC-NT-OF			
M	R	1.1	2.58
RM	No change	0	0
R	RM	0.2	0.46
RTM	No change	0	0
RT	RTM	1.7	3.95
TM	RT	1.4	3.26
T	RT	2.8	6.51
NONE	R	0.9	2.09

Baseline practice	Practices after N_2O credit offered	N_2O reduction	Credit revenue
		Pounds per acre	Dollars per acre
CS-CON-MF			
M	RTM	0.6	1.40
RM	No change	0	0
R	RM	0.2	0.46
RTM	No change	0	0
RT	RM	1.3	3.02
TM	RT	1.6	3.72
T	RT	1.7	3.95
NONE	RTM	0.2	0.46
CS-CON-OF			
M	RTM	0.2	0.46
RM	No change	0	0
R	RM	0.3	0.70
RTM	No change	0	0
RT	RTM	1.2	2.79
TM	RTM	1.1	2.56
T	RT	1.2	2.79
NONE	RTM	0.5	1.16
CS-NT-MF			
M	RT	0.2	0.46
RM	No change	0	0
R	No change	0	0
RTM	No change	0	0
RT	RM	0.8	1.86
TM	RT	1.4	3.26
T	RT	1.4	3.26
NONE	RM	0.5	1.16
CS-NT-OF			
M	R	0.2	0.46
RM	No change	0	0
R	RM	0.2	0.46
RTM	No change	0	0
RT	RTM	1.3	3.02
TM	RTM	1.1	2.56
T	RT	1.4	3.26
NONE	R	0.5	1.16

Note: N = nitrogen. CC = continuous corn, CS = corn-soybeans, CON = conventional till, NT = no-till, MF = manure+inorganic N, OF = inorganic N, M = N incorporate d/injected, R = N rate is less than 40% more than N removed at harvest, T = spring application.

Source: USDA, Economic Research Service.

Farms already meeting both the rate and method criteria will only be able to reduce N_2O emissions by reducing their application rate below recommended rates. The NLEAP modeling indicates only small reductions in N_2O when the application rate is reduced to a level below the criterion rate. This is consistent with field studies that indicate a nonlinear relationship between excessive N application rates and N_2O emissions (Jarecki et al., 2009; McSwiney and Robertson, 2005). Excessive nitrogen inputs accelerate the rate of N_2O emissions. For example, reducing the application rate from the criterion rate to 25 percent below the recommended rate only reduces N_2O by between .2 and 1.3 pounds per acre for the Class A (well-drained) soil in Ohio, depending on the cropping system. Assuming a credit rate of $15 per ton of CO_2 equivalent, this translates into a payment of between $0.46 and $3.02 per acre. These rates are insufficient to cover the 10-percent reduction in corn yields that we assume would occur for such a reduction in N (Bock and Hergert, 1991). Even for smaller N reductions, it is unlikely that revenue from GHG credits would be sufficient to cover the increased risk from cutting N application rates to something close to plant uptake. However, higher offset prices could increase the incentive to cut application rates to reduce N_2O emissions, even when yields might be affected.

When we apply these results to the survey results summarized in table 3.3, we conclude that farmers with treated corn acres meeting the rate, timing, and method criteria or the rate and method criteria (about 42 percent of all corn acres) will not likely participate in a GHG cap-and-trade program that would allow farmers to sell offsets from N_2O reductions. These farms cannot make any management changes to reduce N_2O without reducing overall nitrogen use efficiency, which would violate a market rule. The treatment of such "good stewards" in an emissions trading program is an important policy issue.

The potential revenue from GHG credits produced by reducing N_2O appears to be quite small. In the Ohio example, only a few situations are capable of producing credit revenue of over $5 per acre, assuming a credit price of $15 per ton of CO_2 equivalent (and the results are similar for the other States studied). These rates are less than the rates farmers could receive for nutrient management from EQIP, which is a measure of farmers' willingness to accept payment for the practice (table 4.3). In general, farms overapplying nitrogen and broadcasting fertilizer can produce the largest reductions in N_2O. However, only 8.3 percent of corn acres fall in this category (see table 3.3). While we found that changes in operating costs after changing management are near 0 or even negative in most cases, we did not consider short-term adjustment costs, changes in risk, or the administrative costs of participating in an offset program. In the case of farms that also have animals, we did not consider the cost of moving manure produced on the farm to more acres (to reduce application rates), or of moving excess manure off the farm entirely (Ribaudo et al., 2003)—all of which would reduce farmer participation below the rates estimated here.

One issue of concern is the possibility that reducing N_2O could increase nitrate losses to water. As described in chapters 2 and 3, changes in management could change conditions in the soil so that gaseous forms, such as N_2O, are converted to highly soluble nitrate (NO_3). It might seem that allowing only management changes that do not increase total losses of nitrogen would prevent this, but we found otherwise. In 25 percent of the cases where management changes were made to reduce N_2O, NO_3 losses to water increased, even though total nitrogen emissions fell. This occurred almost exclusively when the rate criterion was already being met and injection/incorporation was adopted as an additional practice. While overall N_2O and total nitrogen losses decreased, water quality worsened. Such an outcome would be a concern in regions trying to address water quality problems, such as the Corn Belt, where corn production is the major source of nitrogen contributing to hypoxia in the Gulf of Mexico. Including these factors in the analysis would likely further reduce the net value to society of producing GHG offsets through N_2O emissions reductions.

Response to Price Changes, and What It Means for an Input Tax

Input prices can influence a farmer's planning. For example, low fertilizer prices can lead to "insurance" applications of fertilizer that reduce overall nitrogen use efficiency. Increases in fertilizer prices relative to other input and output prices through the use of an input tax would likely decrease fertilizer use and reduce the number of acres receiving excessive rates. Several States have levied fertilizer taxes in the past but only at low levels that had little impact on use.

The effectiveness of an input tax in reducing excessive application rates would depend largely on the responsiveness of farmers to changes in nitrogen prices. Data from studies spanning several decades reveal that responses to a price change (known as the price elasticity) can vary widely, depending on the data source and time period covered, the type of econometric methods used to analyze the data, the number of crops covered, and the type of crop to which the nitrogen fertilizer is applied. While no true consensus exists, study findings generally show that nitrogen demand was relatively insensitive to price. Burrell (1989) provides a convenient summary of 14 empirical demand studies through the 1980s. Of those 14 studies, only 4 report elasticities greater than unity. Estimates were generally in the range of -0.20 to -0.70, implying that a 10-percent increase in the price of fertilizer reduced demand by 2 to 7 percent (see, for example, Griliches (1958); Carman (1979); Ray (1982); and Shumway (1983)). Denbaly and Vroomen (1993) use cointegrated and error-corrected models with time series data from 1964 to 1989 to estimate short- and longrun Marshallian elasticities. They report a shortrun Marshallian elasticity of -0.21 and a longrun elasticity of -0.41. Hansen (2004) estimates nitrogen fertilizer demand of farmers in Denmark using an unbalanced panel spanning 1982-91. He concludes that nitrogen demand is similarly insensitive to own-price, with an elasticity of -0.45.

Not all studies found the price elasticity of demand for nitrogen fertilizer to be inelastic. Carman (1979) examines the nitrogen demand in 11 Western States and finds significant State-level variation in elasticities. Statistically significant elasticity estimates in Carman's study range from -0.55 to as large as -1.84. His study shows that demand can vary significantly even within a region. Roberts and Heady (1982) also use annual time-series data from the United States, but spanning 1952-76, and find price elastic demand for nitrogen applied to corn (-1.148). In a study of aggregate fertilizer, Weaver (1983) investigates the demand in just two States, North Dakota and South Dakota, and finds fertilizer demand to be highly elastic, ranging from -1.377 to -2.156. Some evidence suggests that farmers may be becoming more sensitive to changes in fertilizer prices. Using 2001 and 2005 field-level data from ARMS, we estimate a demand elasticity of nitrogen fertilizer of -1.38 for farmers who applied commercial nitrogen fertilizer to corn (app. 3). Stated another way, if the price of nitrogen fertilizer was to rise by 10 percent, farmers would reduce the amount applied by 13.8 percent. At the mean amount of commercial nitrogen, such a change in price would result in a decrease of 18.2 lbs of fertilizer per acre.[5]

Manure can also be used as a source of nitrogen nutrients, usually in conjunction with commercial nitrogen fertilizer. In the ARMS sample, slightly less than a quarter of corn farmers applied manure to the field, and all of them did so in conjunction with commercial nitrogen. When the analysis is expanded to include these farmers, we find a demand elasticity of -0.67; that is, for every 10-percent increase in the price of commercial nitrogen fertilizer, farmers reduce their use of nitrogen (organic and inorganic) by about 7 percent. The results are driven by farmers who use both manure and commercial nitrogen; we find they are relatively less sensitive to the price of commercial nitrogen fertilizer than farmers who apply commercial nitrogen exclusively, which is consistent with the idea that manure and inorganic forms of nutrients are imperfect substitutes. Also, manure management decisions on farms with animals might be driven less by nitrogen prices than by the need to dispose of manure (Ribaudo et al., 2003). The estimates of price elasticity can be used to provide a rough estimate of the tax that would be needed to reduce application rates so that more acres meet the rate criterion. Figure 4.2 displays the distribution of the nitrogen application rates that represent the criterion rate described in chapter 3. In the case of farmers who used commercial nitrogen exclusively, we have estimated an average criterion application rate at 170.8 lbs per acre for production year 2005. Thirty-five percent of the 76 million corn acres treated with nitrogen exceeded their criterion rate (26.7 million acres), and farmers who exceeded their criterion rate had a mean rate of 185.5 pounds per acre. From the distribution depicted in figure 4.3, the concentration of farmers near zero indicates that most of the farmers who applied nitrogen at rates above the criterion rate are situated near the threshold (also seen in table 3.3). In fact, 50 percent of farmers who exceeded the criterion rate exceeded it by 19 pounds per acre or less.

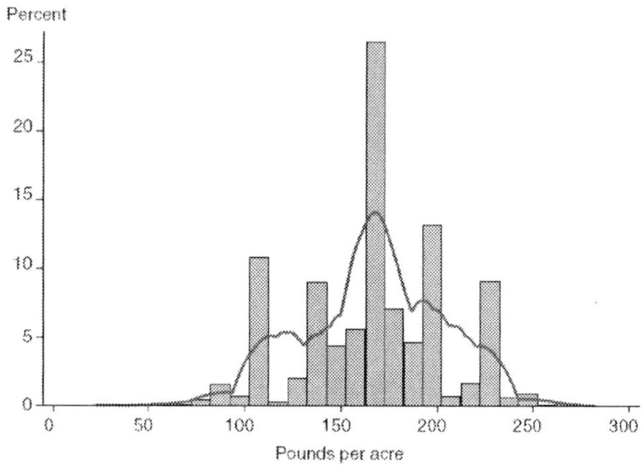

[1]Criterion rate defined as nitrogen removed at harvest plus 40 percent, based on the farmer-stated yield goal.

Note: The kernel density, represented by the smooth line, is an estimate of the continuous density using an Epanechnikov kernel.

Source: USDA, Economic Research Service using USDA's 2005 Agricultural Resource Management Survey.

Figure 4.2. Distribution of criterion rates1 for corn, based on reported expected yield, 2005.

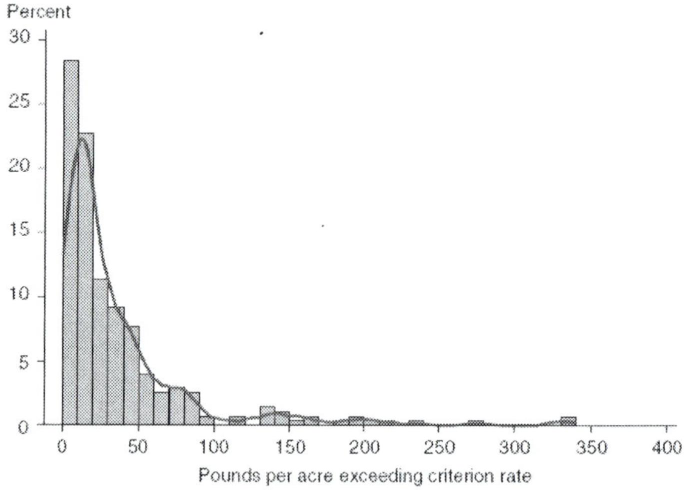

[1] Criterion rate defined as nitrogen removed at harvest plus 40 percent, based on the farmer-stated yield goal.

Note: The kernel density, represented by the smooth line, is an estimate of the continuous density using an Epanechnikov kernel.

Source: USDA, Economic Research Service using USDA's 2005 Agricultural Resource Management Survey.

Figure 4.3. Distribution of nitrogen fertilizer applied to corn that exceeded the criterion rate,[1] 2005.

Table 4.7 provides a summary of the input tax needed to reduce the excess use of nitrogen by farmers who exceed their criterion rate, evaluated for differing levels of demand elasticity. From the table it is evident that the more elastic the demand, the less the price must change to reduce excessive application rates. A highly inelastic demand for nitrogen, for example -0.20, would require more than a 50-percent increase in the price to achieve a 50-percent reduction in excess application. To achieve a reduction of 75 percent, the price would have to more than double.

Based on our estimated elasticity of -1.38, if an input tax increased the price of nitrogen by 7.4 percent, 50 percent (about 13.4 million acres) of the 26.7 million overtreated acres would then meet the rate criterion. Seventy-five percent of heavy nitrogen users exceed the criterion rate by 43.4 pounds per acre or less; thus, raising the price of nitrogen by 17 percent would reduce cropland exceeding the criterion rate by 20 million acres. For context, consider the mean price of nitrogen fertilizer in 2005 was 33 cents per pound; therefore, a 7.4-percent change in the price equates to slightly more than 2.4 cents per pound, and a 17-percent change equates to less than 6 cents per lb.

As a policy instrument, a tax on inputs has some desirable characteristics as well as some well-known drawbacks. First, a tax gives farmers flexibility in how they reduce emissions. Farmers face heterogeneous costs, and a tax enables farmers to tailor their input responses (nitrogen abatement) accordingly (Ribaudo et al., 1999). In the case of nitrogen, an input tax directly affects the farmer's decision that has the largest impact on nitrogen losses to the environment. It would also encourage a farmer to manage nitrogen more carefully, which could lead to appropriate timing and method of application. A tax does not require monitoring or enforcement, unlike a regulation. It can also be easily adjusted if policy goals are not met or exceeded. Another advantage of an input tax is that it raises revenue while reducing application rates. The revenue could be used to reduce the tax burden of crop producers through a system of lump-sum rebates to those producers who improve nitrogen use efficiency. Revenue can also be used to remedy damages caused by nitrogen losses.

Table 4.7. Fertilizer price increases needed to reduce excess nitrogen† applications by 50 percent and 75 percent

	Reduce excess nitrogen application by:			
	50 percent		75 percent	
Elasticity of nitrogen fertilizer demand	Necessary price change	Tax	Necessary price change	Tax
	Percent	*Dollars*	*Percent*	*Dollars*
-0.20	51.2	0.169	117.0	0.386
-0.50	20.5	0.068	46.0	0.154
-0.70	14.6	0.048	33.4	0.110
-1.00	10.2	0.034	23.4	0.077
-1.38	7.4	0.024	17.0	0.056

Note: † Excess nitrogen application is defined as rate exceeding 40 percent more than nitrogen removed at harvest (see chapter 3).
Source: USDA, Economic Research Service.

A tax on an input also has drawbacks. An input tax makes no distinction between whether fertilizer is in excess or not. A tax on nitrogen may also encourage increased use of untaxed manure, resulting in no discernable change in nitrogen applications where manure is readily available.

The question of who bears the burden of the tax, also known as the incidence, can have notable distributional consequences. Statutorily, the incidence of the tax could fall on the wholesaler or retailer of nitrogen fertilizer; however, the true, or economic, incidence is likely to be shared with the farmer. How much so is an empirical question that relies on the relative sensitivity of farmers to the price change, as well as the elasticity of the supply of nitrogen: the more sensitive a farmer's demand for nitrogen is, the less of a burden he or she will bear, all else equal. The supply of nitrogen fertilizer is projected to more than meet the demand over the near term; therefore, the standard assumption is that the burden of the excise tax would be considerably shifted to the consumer of the good, in this case the farmer (Fullerton and Metcalf, 2002; FAO, 2008). While corn production in the United States accounts for 39 percent of the world's total corn production, the ability of U.S. farmers to pass along the cost of the tax will depend on the relative elasticities of supply and demand for corn (USDA, FSA, 2011). While a factor tax on nitrogen may improve welfare from society's point of view, ultimately, the tax will change the functional distribution of income. The distributional impact may be mitigated if revenues raised by the tax are returned to the farmer in some manner, for example, by supporting other conservation activities.

Nitrogen Compliance

Compliance provisions require farmers to meet some minimum standard of environmental protection on environmentally sensitive land as a condition for eligibility for many Federal farm program benefits, including conservation and commodity program payments. Under current compliance requirements, farm program eligibility could be denied to producers who:

- Fail to implement and maintain an NRCS-approved soil conservation system on highly erodible land (HEL) (Conservation compliance)
- Convert HEL grasslands to crop production without applying an approved soil conservation system (Sodbuster)
- Convert a wetland to crop production (Swampbuster)

Evidence suggests that the current compliance provisions have contributed to a reduction in soil erosion and discouraged the conversions of noncropped HEL land and wetlands to cropland (Claassen et al., 2004). A possible extension of the provisions could include nutrient management.

Crop producers are a major source of nitrogen. Assessments of the potential efficacy of compliance must consider two key questions:

- To what extent do crop producers who have the greatest potential for reducing nitrogen emissions also participate in farm programs?
- Are government payments to these producers large enough to encourage broad adoption of practices that improve nitrogen use efficiency and reduce nitrogen emissions?

Claassen et al. (2004) estimate that 75 percent or more of cropland acres with medium, high, or very high potential for nitrogen leaching or runoff are located on farms that receive government payments. We used data from the 2005 ARMS corn survey to estimate government payments received by corn producers.[6] We looked at all treated corn acres, as compliance provides an incentive both for farmers already practicing good nitrogen management and willing to continue and for farmers not using nitrogen BMPs and willing to adopt them. Over 97 percent of corn acres receive government payments, averaging $51.39 per acre. This average is higher than our estimated costs of improving NUE or of adopting NRCS practices. Eighty-eight percent of treated corn acres receive government payments in excess of $27 per acre per year, which is more than the average EQIP payments for nutrient management or waste use. (Note that for corn acres that are highly erodible and subject to conservation compliance, it is the sum of erosion control and nitrogen management costs that would be considered by the farmer.)

A drawback of compliance is that the strength of the incentive is dependent on the level of government payments. Current events present a good example. Direct government payments have been reduced by about 50 percent between 2005 and 2009 due to a number of factors, including higher crop prices and smaller disaster payments (USDA, ERS, 2010). Assuming that average per acre payments to corn producers were reduced by the same percentage, the average estimated cost of the more expensive nitrogen management practices, such as waste utilization, would be greater than the program benefit. Compliance would not be an effective tool in this case. The point is that program payments can vary greatly, making compliance an unpredictable policy instrument.

Regulation

Another policy approach for improving NUE is to legally require farms to adopt and implement particular management practices. Such an approach would be a major change in the way most of agriculture is treated under current environmental laws. With few exceptions, agricultural operations are exempt from regulation under the Clean Water Act and Clean Air Act. A number of arguments have been used as justification. First, agriculture is so diverse across the United States that the conventional regulatory approach of applying uniform standards is impractical (Nanda, 2006). Second, due to the nonpoint nature of agricultural pollution, individual polluters cannot be identified except at great cost. Regulation can conceptually be placed on a continuum between performance standards and design standards (Ribaudo et al., 1999). Performance standards directly regulate emissions. Design standards dictate how producers manage their operations, including practices that should not be used and/or BMPs that should be adopted. Because of the nonpoint nature of agricultural pollution, design standards are the only practical approach for addressing nitrogen losses.

One approach would be to require that farmers adopt specific BMPs to improve their nitrogen use efficiency. Generally, a practice-based regulation is inefficient because it requires producers to adopt the same practice, whether it is appropriate for their particular farm or not. It may be more effective to define BMPs locally so as to allow flexibility and to account for agriculture's heterogeneous nature. For example, a nitrogen management plan is a flexible practice that is based on a farmer's resources and cropping system. However, farmers may fail to implement the plan properly. The effectiveness of a regulation therefore requires effective inspection and enforcement by a resource management agency. Implementation costs would likely be high. Several States, such as Nebraska and Maryland, have required farmers in particularly vulnerable areas to adopt specific nutrient management practices to protect ground or surface water (Ribaudo, 2009).

One of the few segments of the agricultural sector that has been subjected to regulatory environmental measures at the national level is animal feeding operations, reflecting heightened concern over pollution from animal waste from the largest operations (USDA-EPA, 1999). Manure is estimated to be a source of about 17 percent of nitrogen entering U.S. waters (Smith et al., 1997).

Clean Water Act regulations now require that animal feeding operations designated as Concentrated Animal Feeding Operations, or CAFOs, and needing a National Pollutant Discharge Elimination System (NPDES) permit (those CAFOs that discharge or propose to discharge to surface waters), develop and implement a nutrient management plan to cover fields that receive manure.

Such a plan, which would meet NRCS standards, sets a limit on the amount of nutrients that can be applied per acre of land and specifies erosion control measures to prevent the loss of sediment and nutrients. Also under the new regulations, CAFOs that are not required to have an NPDES permit but that wish to claim the storm water exemption (the provision in the Clean Water Act that exempts field practices from requiring a discharge permit) for runoff from fields must develop and implement a nutrient management plan to demonstrate that due care is being taken to minimize polluted runoff from fields receiving manure. If a waterway becomes polluted with animal waste from field runoff and a CAFO does not have an approved nutrient management plan, this would be a violation of the Clean Water Act. This approach sets a level of expected stewardship, namely the implementation of a nutrient management plan. Requiring not just CAFOs but all animal feeding operations to adopt nutrient management plans would be costly. ERS estimates that reductions in net returns in the livestock and poultry sector would be about $1.4 billion per year, and national economic welfare for producers and consumers would decline almost $2 billion per year (Ribaudo et al., 2003). The benefit would be improved air and water quality. Targeting the regulatory approach only to those operations most susceptible to pollution problems would lower the overall costs.

Implications for Nitrogen Management Policies

Nitrogen is critical for producing abundant food and generating high net returns to producers, yet it has wide-ranging environmental impacts across land, water, and the atmosphere. More careful management that reduces environmental losses would address a number of environmental issues, such as hypoxia in coastal estuaries and bays, the potential for global warming, and nutrient enrichment of terrestrial ecosystems. Policymakers have a number of tools at their disposal, each with its own strengths and weaknesses (table 5.1). No one policy approach can be considered "best," and a concerted effort to address the nation's nitrogen problems will likely require a solution comprising a mix of policies. Our analysis provides some guidance on determining which sectors of agriculture are most in need of improved management, what are the potential pitfalls, and how might the different policies be orchestrated in an overall policy framework.

Reducing Application Rates as a Priority Policy Goal

Reducing the application of nitrogen fertilizers appears to be the most effective BMP for reducing the emission of nitrogen into the environment. Based on the literature, and confirmed by our NLEAP modeling, reducing application rates is the one BMP that reduces *all* forms of reactive nitrogen, even when the timing and method of application are not ideal. Improving timing or method of application alone could increase one type of reactive nitrogen (transmitted to the atmosphere, groundwater, or surface water) while still reducing total nitrogen emissions. Reducing the application rate is therefore conducive to an ecosystem approach to management that provides protection to all ecosystem services and functions. Improving rate, timing, and method of nitrogen application would produce the greatest environmental benefits. Reducing application rates that are agronomically excessive may increase the perceived risk of reduced yields. Farmers often use nitrogen fertilizer to manage the downside risk due to uncertain weather and soil nitrogen. Research on how farmers view risk and how they might respond to an incentive payment for reducing application rates, coupled with the use of a risk management instrument, could result in the development of a more effective approach for reducing nitrogen in the environment. Revenue or yield insurance policies could be offered to protect the income of farmers who adopt conservation measures that improve nitrogen use efficiency but may decrease yields because of nitrogen insufficiency stemming from unfavorable weather conditions. Findings from other studies suggest that insurance will likely lead to reductions in nitrogen fertilizer applications, but by how much is uncertain (see Babcock and Hennessy, 1996; Mishra et al., 2005; Smith and Goodwin, 1996).

Corn Is the Most Important Crop for Addressing Nitrogen-Related Environmental Issues

Corn is the most widely planted crop in the United States and the most intensive user of nitrogen. In 2006, corn accounted for an estimated 65 percent of the total quantity of nitrogen applied to major U.S. field crops. Corn also accounted for half of all nitrogen-treated crop acres that were not meeting the rate, timing, or method of application criteria used in this analysis to define acceptable nitrogen management.

Table 5.1. Summary of policy instruments for improving nitrogen use efficiency

	Policy instrument					
Characteristics	Input tax	Information	Financial incentives	Compliance	Emissions market	Regulation
Strength of incentive	Depends on level of tax and price elasticity of demand.	A farmer will take action only if management practice improves profi ts.	Depends on level of subsidy.	Depends on level of Government Program payments subject to compliance.	Depends on level of demand from regulated sectors.	Strong.
Acres covered	Covers all acres that are treated with commercial nitrogen.	No guarantee that acres in need of treatment will be addressed.	No guarantee that acres most in need of treatment will be addressed.	May not cover all acres.	May be limited by geographic scope of market and baseline rules.	Can cover all acres that use commercial nitrogen or animal waste.
Targets problem	Directly addresses application rate, but not timing and method. Also, does not address application of animal waste.	Information can be targeted to specific problems.	Incentives can be targeted to specific practices and regions. However, important to consider potential environmental tradeoffs.	Strength of incentive may not be correlated with acres most in need of treatment.	Generally limited to one pollutant and not overall nitrogen use efficiency. Environmental tradeoffs a potential problem.	Can target all aspects of nutrient management. However, important to consider potential environmental tradeoffs.
Flexibility	Very fl exible – farmers can adjust in the most cost effective way.	Flexible –farmers act on information that is benefi cial to them.	Practice-based incentives are less fl exible than incentives on environmental performance.	Flexibility depends on how provisions are defi ned.	Can be fl exible, but depends on market rules.	Limited flexibility, as regulations generally require specific practices.

Table 5.1. (Continued)

	Policy instrument					
Characteristics	Input tax	Information	Financial incentives	Compliance	Emissions market	Regulation
Implementation costs	Easy to implement, and generates revenues that can be used to reduce economic impacts for farmers who make improvements.	Requires research and extension outreach.	High costs to taxpayers.	Enforcement costs may be high.	Transactions costs can be very high.	Enforcement costs can be high.

Land used to grow corn accounted for the largest share of treated acres that had tile drainage in 2006. Although tile drains improve yields, they also increase the amount of nitrogen that is lost to surface water. Tiled corn cropland not meeting all three nitrogen management criteria would be a prime target for policies for improving nitrogen use efficiency.

In addition, recent demand pressures due to the biofuels mandate, as well as increasing international demand for feed grains, suggests that corn acreage and the intensity of corn production are likely to increase. Together, these factors increase the importance of raising the NUE in corn production in the United States, especially on farms that raise livestock and apply manure to their fields.

Which Policy Is Best?

This analysis provides some guidance on how different policies might be orchestrated in an overall policy framework. The current approach to improving nutrient management on cropland has relied primarily on financial incentives and information. While years of financial and technical assistance have resulted in some progress, operators of over 65 percent of U.S. cropland are still not implementing nitrogen BMPs. Higher payment rates would encourage more producers to adopt practices that improve nitrogen use efficiency, but the cost to taxpayers may be substantial. The level of financial assistance that would be required to entice all farmers with cropland acres needing improved management to enroll in a program would likely consume most of the budget for EQIP. While nitrogen management is an important conservation goal, EQIP and other USDA conservation programs address a host of other issues. Any elevation of nitrogen management as a priority for EQIP may result in fewer resources for other conservation issues.

Emissions markets, such as those for water quality or greenhouse gases, could be a source of financial support for improving nitrogen use efficiency. Markets for agricultural offsets shift the financial burden away from taxpayers to regulated sectors of the economy. While emissions markets are receiving much interest in efforts to improve water quality and to reduce greenhouse gas emissions, their role in improving nutrient management on all acres needing improvement is probably limited. Emissions markets generally target particular geographic areas or particular practices, potentially limiting the number of acres that might be affected. Market rules designed to ensure the "additionality" of offsets by setting baselines consistent with a high level of management may limit participation by farmers not using BMPs, even though a market would benefit by their participation. In addition, the nonpoint source nature of nitrogen emissions from agriculture greatly complicates the design of markets and raises transactions costs.

If voluntary financial assistance programs or emissions markets are limited in their ability to improve nitrogen management across all crop acres, what other approaches might achieve improved nitrogen use efficiency at least cost? The alternative approaches all result in increased costs for farmers. In theory, cost-effective policy instruments target the problem, are flexible, are easy to implement (low transactions costs), and limit costs to both farmers and government. A tax on nitrogen fertilizer would provide an incentive to all users to manage commercial nitrogen more carefully. If farmers are responsive to price, then this instrument may be an effective means of reducing nitrogen losses. Our assessment of farmer price responsiveness indicates that a relatively low tax may pay high environmental dividends. However, if farmers are as unresponsive to nitrogen prices as generally reported in the literature, a substantially higher tax would be necessary to obtain the same environmental benefits. The burden on farmers would be substantial. Another drawback of an input tax is that a tax would also be paid on applications that are not excessive. A tax only on emissions would be far more efficient, but such a tax is not practical since emissions cannot be observed or easily measured. Finally, some means of addressing the application of animal waste would have to be found, as a fertilizer tax would likely encourage the substitution of manure for commercial nitrogen.

A nutrient management plan is an inherently flexible management practice that is strongly encouraged by USDA but only required for animal feeding operations that are designated as CAFOs. Requiring that all users of nitrogen inputs (commercial and manure) develop and implement a nutrient management plan would be a major change in the way the environmental performance of agriculture is managed. The costs to crop farmers of implementing a nutrient management plan may not be high, except for those managing large amounts of manure produced on the farm. However, many aspects of a nutrient management plan, such as application rate, are difficult to observe, making enforcement difficult and costly.

Enforcement costs could also be high for a compliance approach to getting farmers to adopt nutrient management plans. The effectiveness of compliance would depend on the level of program payments received by farmers and a coincidence of the incentive with those crop acres most in need of improved management. A large share of crop acres in need of treatment receives high levels of government program payments. While the incentive level in 2005 was quite high, program payments have declined in recent years as crop prices have risen. Continued high prices and general concerns about Federal budget outlays may limit the strength of a compliance-type policy instrument unless it is linked to a broader suite of payments than current compliance requirements.

Improving nitrogen use efficiency reduces the amount of emissions from cropland but does not eliminate them. In areas where even small levels of emissions could cause environmental problems, offsite filtering could supplement onfield management. The government currently provides financial incentives for creating and preserving wetlands and vegetative filter strips. Though funds are not allocated solely for nitrogen capture and removal, there may be reasons to do so. An economic comparison of the two types of filters suggests that wetlands can be much more cost effective at removing nitrogen than filter strips. While our analysis found that the cost of establishing a wetland is greater than the cost of establishing filter strips, annual nitrogen removal rates are several times greater for wetlands. Filter strips may also be rendered ineffective where tile drains are present, while wetlands can be strategically positioned in the landscape to filter drainage coming from tiled fields. Wetlands also produce a number of other desirable ecosystem services, such as wildlife habitat. Filter strips, however, can be established in landscape settings where wetlands cannot. The choice will depend on geography, soil, and hydrologic conditions.

While one single policy instrument does not emerge as a clearly superior approach to improving NUE across all cropland, a role can be seen for each. Financial assistance could be made available to those producers wanting to voluntarily improve nutrient management and to install vegetative filters or resore wetlands. Since commodity programs are important to farmers, compliance can provide some incentive for those receiving program payments. The level of incentive may vary from year to year, but it may be effective for some farmers. Finally, in regions where nitrogen-related pollution is of particular concern, such as the Chesapeake Bay watershed and the watersheds contributing nitrogen to the Gulf of Mexico, a regulatory backstop could be a measure of last resort for those unwilling to voluntarily adopt nitrogen BMPs.

Information Supports All Policies

Information about the environmental and economic performance of improved nitrogen management practices supports all policies aimed at improving NUE. Reliable, timely information on soil and plant nitrogen reduces one source of uncertainty that tends to encourage overapplication of nitrogen. Our research supports previous findings that testing for nitrogen available in the soil and contained in crops may result in lower application rates. Information from testing can be incorporated into an adaptive management framework, where a farmer evaluates his practices from the previous year (or even at the start of the current growing season) to assess what options may be available to improve nutrient management while sustaining yields and reducing nutrient losses to the environment. So, whether farmers are considering best nitrogen management practices due to regulation, taxes, or financial incentives, information on how to conduct and interpret nitrogen tests and how to successfully implement new practices can reduce the overall costs and increase adoption rates.

Potential Tradeoffs Are an Important Consideration

Reactive nitrogen is easily converted to forms that are readily transported by hydrologic and atmospheric processes. Therefore, focusing strictly on one issue, such as nitrate leaching, could lead to increased emissions of other nitrogen compounds, such as nitrous oxide to the atmosphere, if nitrogen's characteristics are ignored. Even when total nitrogen emissions are reduced by a policy, emissions of one or more nitrogen compounds might increase and degrade environmental quality. This effect was predicted in the case of the market for nitrous oxide offsets—farmers reduced total emissions but increased nitrogen losses to water. These tradeoffs often depend on soils and cropping practices, so it is difficult to develop general "rules of thumb," other than recommending that a holistic approach to management that considers potential environmental tradeoffs be adopted. Reducing nitrogen application rates is the easiest and most effective way to reduce all forms of reactive nitrogen.

References

Abt Associates. 2000. *Air Quality Impacts of Livestock Waste*. Report prepared for U.S. Environmental Protection Agency, Office of Policy.

Alley, M.M., M.E. Martz, Jr., P.H. Davis, and J.L. Hammons. 2009. *Nitrogen and Phosphorus Fertilization of Corn*. Publ. 424-027. Virginia

Cooperative Ext., Virginia Polytechnic Institute and State University. (Available online at www.pubs.ext.vt.edu/424/424-027/424-027.html.) (Verified June 3, 2010.)

Alva, A.K., S. Paramasivam, K. Sajwan, A. Fares, J.A. Delgado, and D. Mattos. 2005. "Nitrogen and Irrigation Management Practices To Improve Nitrogen Uptake Efficiency and Minimize Leaching," *Journal of Crop Improvement* 15(2):369-420.

Angier, J.T., G.W. McCarty, T.J. Gish, and C.S. Daughtry. 2001. "Impact of a First-Order Riparian Zone on Nitrogen Removal and Export From an Agricultural Ecosystem," *TheScientific World* 1(Suppl 2):642-651.

Babcock, B.A. 1992. "The Effects of Uncertainty on Optimal Nitrogen Applications," *Review of Agricultural Economics* 14(2):271-280.

Babcock, B.A., and A.M. Blackmer. 1992. "The Value of Reducing Temporal Input Nonuniformities," *Journal of Agricultural and Resource Economics* 17:355-347.

Baker, J. 2001. "Limitations of Improved Nitrogen Management to Reduce Nitrate Leaching and Increase Use Efficiency," in Optimizing Nitrogen Management in Food and Energy Production and Environmental Protection: Proceedings of the 2nd International Nitrogen Conference on Science and Policy, *TheScientificWorld* 1(S2):10-16.

Bausch, W.C., and J.A. Delgado. 2003. "Ground-Based Sensing of Plant Nitrogen Status in Irrigated Corn To Improve Nitrogen Management," in T. Van Toai, D. Major, M. McDonald, J. Schepers, and L. Tarpley (eds.), *Digital Imaging and Spectral Techniques: Applications to Precision Agriculture and Crop Physiology*, ASA Spec. Publ. 66, American Society of Agronomy, Crop Science Society of America, Soil Science Society of America. pp. 151-163.

Beckie, H.J., A.P. Moulin, C.A. Campbell, and S.A. Brandt. 1995. "Testing Effectiveness of Four Simulation Models for Estimating Nitrates and Water in Two Soils," *Canadian Journal of Soil Sciences* 75(1):135-143.

Beegle, D.B., and P.T. Durst. 2003. *Nitrogen Fertilization of Corn*, Agronomy Facts 12, College of Agricultural Sciences Cooperative Extension, The Pennsylvania State University. (Available online at http://export.nbii.gov/xml/web-resources/xmlfiles)

Bishop, C.P., C.R. Shumway, and P.R. Wandschneider. 2010. "Agent Heterogeneity in Adoption of Anaerobic Digestion Technology: Integrating Economic, Diffusion, and Behavioral Innovation Theories," *Land Economics* 86(5):585-608.

Bock, B.R., and W. Hergert. 1991. "Fertilizer Nitrogen Management," in H. Follet et al. (eds.), *Managing Nitrogen for Groundwater Quality and Farm Profitability*, Soil Science Society of America, pp. 140-164.

Bosch, D.J., Z.L. Cook, and K.O. Fuglie. 1995. "Voluntary Versus Mandatory Agricultural Policies To Protect Water Quality: Adoption of Nitrogen Testing in Nebraska," *Review of Agricultural Economics* 17:13-24.

Breetz, H.L., K. Fisher-Vander, L. Garzon, H. Jacops, K. Kroetz, and R. Terry. 2004. *Water Quality Trading and Offset Initiatives in the U.S.: A Comprehensive Survey,* Dartmouth College. (Available online at http://www.dartmouth.edu/~kfv/waterqualitytradingdatabase.pdf)

Bricker, S., B. Longstaff, W. Dennison, A. Jones, K. Boicourt, C. Wicks, and J. Woerner. 2007. *Effects of Nutrient Enrichment in the Nation's Estuaries: A Decade of Change*, NOAA Coastal Ocean Program Decision Analysis Series No. 26, National Centers for Coastal Ocean Science.

Burrell, A. 1989. "The Demand for Fertilizer in the United Kingdom," *Journal of Agricultural Economics* 40:1-20.

Camacho, R. 1992. *Chesapeake Bay Program Nutrient Reduction Strategy Reevaluation. Financial Cost Effectiveness of Point and Nonpoint Source Nutrient Reduction Technologies in the Chesapeake Bay Basin*, Interstate Commission on the Potomac River Basin Report No. 92-4.

Carman, H.F. 1979. "The Demand for Nitrogen, Phosphorous, and Potash Fertilizer Nutrients in the Western United States," *Western Journal of Agricultural Economics* 4(1):23-32.

Cassman, K.G., A. Dobermann, and D.T. Walters. 2002. "Agroecosystems, Nitrogen-Use Efficiency, and Nitrogen Management," *Ambio* 31(2):132-140.

Chouinard, H.H., T. Paterson, P.R. Wandschneider, and A.M. Ohler. 2008. "Will Farmers Trade Profits for Stewardship? Heterogeneous Motivations for Farm Practice Selection," *Land Economics* 84 (1): 66-82.

Claassen, R., V. Breneman, S. Bucholtz, A. Cattaneo, R. Johansson, and M. Morehart. 2004. *Environmental Compliance in U.S. Agricultural Policy: Past Performance and Future Potential*, Agricultural Economic Report No. 832, U.S. Department of Agriculture, Economic Research Service. (Available online at www.ers.usda.gov/publications/aer832)

Cooper, J.C., and R.W. Keim. 1996. "Payments To Encourage Farmer Adoption of Water Quality Protection Practices," *American Journal of Agricultural Economics* 78(1): 54-64.

Cowling, E., J. Galloway, C. Furiness, and J.W. Erisman et al. 2002. "Optimizing Nitrogen Management and Energy Production and

Environmental Protection." Presented at the Second International Nitrogen Conference. Bolger Center, Potomac, MD, October 14-18, 2001. (Available at www.initrogen.org/fileadmin/user_upload/Second_N_Conf_Report. pdf. (Verified May 10, 2010).

Crumpton, W. G., D. A. Kovacic, D.L. Hey, and J.A. Kostel. 2008. "Potential of Restored and Constructed Wetlands To Reduce Nutrient Export From Agricultural Watershed in the Corn Belt," in Final Report: Gulf Hypoxia and Local Water Quality Concerns Workshop, American Society of Agricultural and Biological Engineers.

Crutchfield, S.R., P.M. Feather, and D.R. Hellerstein. 1995. *The Benefits of Protecting Rural Water Quality: An Empirical Analysis*. Agricultural Economic Report No. 701, U.S. Department of Agriculture, Economic Research Service, January. (Available online at www.ers.usda.gov/publications/aer701)

David, M.B., L.E. Drinkwater, and G.E. McIsaac. 2010. "Sources of Nitrate Yields in the Mississippi River Basin," *Journal of Environmental Quality* 39:1657-1667.

Davis, J.G., K.V. Iversen, and M.F. Vigil. 2002. "Nutrient Variability in Manures: Implications for Sampling and Regional Database Creation," *Journal of Soil and Water Conservation* 57:473-478.

Delgado, J.A. 2001. "Use of Simulations for Evaluation of Best Management Practices on Irrigated Cropping Systems," pp. 355-381, in M.J. Shaffer, L. Ma, and S. Hansen (eds.), *Modeling Carbon and Nitrogen Dynamics for Soil Management*, Boca Raton, FL: Lewis Publishers.

Delgado, J.A., and M.J. Shaffer. 2008. "Nitrogen Management Modeling Techniques: Assessing Cropping Systems/Landscape Combinations," p. 539-570, in R.F. Follett and J.L. Hatfield (eds.), *Nitrogen in the Environment: Sources, Problems and Management*. New York: Elsevier Science.

Delgado, J.A., and W. Bausch. 2005. "Potential Use of Precision Conservation Techniques To Reduce Nitrate Leaching in Irrigated Crops," *Journal of Soil and Water Conservation* 60:379-387.

Delgado, J.A., A.R. Mosier, R.H. Follett, R.F. Follett, D.G. Westfall, L.K. Klemedtsson, and J. Vermeulen. 1996. "Effects of N Management on N_2O and CH_4 Fluxes and ^{15}N Recovery," *Nutrient Cycling in Agroecosystems* 46:127-134.

Delgado, J.A., C.M. Gross, H. Lal, H. Cover, P. Gagliardi, S.P. McKinney, E. Hesketh, and M.J. Shaffer. 2010a. "A New GIS Nitrogen Trading Tool

Concept for Conservation and Reduction of Reactive Nitrogen Losses to the Environment," *Advances in Agronomy* 105:117-171.

Delgado, J.A., M. Shaffer, C. Hu, R. Lavado, J. Cueto-Wong, P. Joosse, D. Sotomayor, W. Colon, R. Follett, S. DelGrosso, X. Li, and H. Rimski-Korsavok. 2008a. "An Index Approach to Assess Nitrogen Losses to the Environment," *Ecological Engineering* 32:108–120.

Delgado, J.A., M.J. Shaffer, H. Lal, S.P. McKinney, C.M. Gross, and H. Cover. 2008b. "Assessment of Nitrogen Losses to the Environment With a Nitrogen Trading Tool (NTT)," *Computers and Electronics in Agriculture* 63:193-206.

Delgado, J.A., R. Khosla, W.C. Bausch, D.G. Westfall, and D. Inman. 2005. "Nitrogen Fertilizer Management Based on Site-Specific Management Zones Reduce Potential for NO_3-N Leaching," *Journal of Soil and Water Conservation* 60:402–410.

Delgado, J.A., R.F. Follett, and M.J. Shaffer. 2000. "Simulation of NO_3-N Dynamics for Cropping Systems With Different Rooting Depths," *Soil Science Society of America Journal* 64:1050-1054.

Delgado, J.A., R. Ristau, M.A. Dillon, H.R. Duke, A. Stuebe, R.F. Follett, M.J. Shaffer, R.R. Riggenbach, R.T. Sparks, A. Thompson, L.M. Kawanabe, A. Kunugi et al. 2001. "Use of Innovative Tools To Increase Nitrogen Use Efficiency and Protect Environmental Quality in Crop Rotations," *Communications in Soil Science and Plant Analysis* 32:1321-1354.

Delgado, J.A., S.J. Del Grosso, and S.M. Ogle. 2010b. "^{15}N Isotopic Crop Residue Cycling Studies and Modeling Suggest That IPCC Methodologies To Assess Residue Contributions to N_2O-N Emissions Should Be Reevaluated," *Nutrient Cycling in Agroecosystems* 86:383–390.

Denbaly, M., and H. Vroomen. 1993. "Dynamic Fertilizer Nutrient Demands for Corn: A Cointegrated and Error-Correcting System," *American Journal of Agricultural Economics* 75(1): 203-209.

Dobermann, A., R. Ferguson, G. Hergert, C. Shapiro, D. Tarkalson, D.T. Walters, and C. Wortmann. 2006. "Nitrogen Response in High-Yielding Corn Systems of Nebraska," in Proceedings of the Great Plains Soil Fertility Conference, Denver, CO, March 7-8, Vol. 11, pp. 50-59, Potash and Phosphate Institute, Brookings, SD.

Dodds, W.K., W.W. Bouska, J.L. Eitzmann, T.J. Pilger, K.L. Pitts, A.J. Riley, J.T. Schloesser, and D.J. Thornbrugh. 2009. "Eutrophication of U.S. Freshwaters: Analysis of Potential Economic Damages," *Environmental Science and Technology* 43(1):12-19.

Doerge, T.A., R.L. Roth, B.R. Gardner (eds.). 1991. *Nitrogen Fertilizer Management in Arizona*, University of Arizona, College of Agriculture.

Dosskey, M.G., D.E. Eisenhauer, and M.J. Helmers. 2005. "Establishing Conservation Buffers Using Precision Information," *Journal of Soil and Water Conservation* 62:349-354.

Dosskey, M.G., M.J. Helmers, and D.E. Eisenhauer. 2007. "An Approach for Using Soil Surveys To Guide the Placement of Water Quality Buffers," *Journal of Soil and Water Conservation* 61:344-354.

Drury, C.F., W.D. Reynolds, C.S. Tan, T.W. Welacky, W. Calder, and N.B. McLaughlin. 2006. "Emissions of Nitrous Oxide and Carbon Dioxide: Influence of Tillage Type and Nitrogen Placement Depth," *Soil Science Society of America Journal* 70(2):570-581.

Dubrovsky, N.M., K.R. Burow, G.M. Clark, J.A.M. Gronberg, P.A. Hamilton, K.J. Hitt, D.K. Mueller, M.D. Munn, L.J. Puckett, B.T. Nolan, M.G. Rupert, T.M. Short, N.E. Spahr, L.A. Sprague, and W. G. Wilbur. 2010. *The Quality of Our Nation's Waters - Nutrients in the Nation's Streams and Groundwater, 1992-2004*. Circular-1350, U.S. Geological Survey.

Eghball, B., B.J. Wienhold, J.E. Gilley, and R.A. Eigenberg. 2002. "Mineralization of Manure Nutrients," *Journal of Soil and Water Conservation* 57:470-473.

Espinoza, L., and J. Ross. 2008. "Fertilization and Liming," pp. 23-27, in L. Espinoza and J. Ross (eds.), *Corn Production Handbook*. Cooperative Extension Service, University of Arkansas, Little Rock, AR. (Available online at www.uaex.edu/Other_Areas/ publications/HTML/MP-437.asp.) (Verified June 3, 2010.)

Flessa, H., and F. Beese. 2000. "Laboratory Estimates of Trace Gas Emissions Following Surface Application and Injection of Cattle Slurry," *Journal of Environmental Quality* 29: 262-268.

Follett, J.R., R.F. Follett, and W.C. Herz. 2010. "Environmental and Human Impacts of Reactive Nitrogen," in J.A. Delgado and R.F. Follett (eds.), *Advances in Nitrogen Management for Water Quality*, Soil and Water Conservation Society.

Food and Agriculture Organization of the United Nations (FAO). 2008. *Current World Fertilizer Trends and Outlook to 2012*, Rome.

Fox, R.H., W.P. Piekielek, and K.E. Macneal. 1996. "Estimating Ammonia Volatilization Losses From Urea Fertilizers Using a Simplified Micrometerological Sampler," *Soil Science Society of America Journal* 60:596-601.

Freney, J.R., J.R. Simpson, and O.T. Denmead. 1981. "Ammonia Volatilization," *Ecological Bulletin* 33:291-302.

Freney, J.R., M.B. Peoples, and A.R. Mosier. 1995. *Efficient Use of Fertilizer Nitrogen by Crops*, Extension Bulletin 414, Food & Fertilizer Technology Center, Taipei, Taiwan. (Available online at www.agnet.org/library/eb/414) (Accessed December 29, 2009.)

Fullerton, D., and G.E. Metcalf. 2002. "Chapter 26 Tax Incidence," *Handbook of Public Economics* 4:1787-1872.

Galloway, J.N., A.R. Townsend, J.W. Erisman, M. Bekunda, Z. Cai, J.R. Freney, L.A. Martinelli, S.P. Seitzinger, and M.A. Suttan. 2008. "Transformation of the Nitrogen Cycle: Recent Trends, Questions, and Potential Solutions," *Science 320*(May 16):889-892.

Galloway, J.N., J.D. Aber, J.W. Erisman, S.P. Seitzinger, R.W. Howarth, E.B. Cowling, and B.J. Cosby, 2003, "The Nitrogen Cascade," *BioScience* 53(4): 341-356.

Ghosh, G., Ribaudo, M., and J. Shortle. 2011. "Do Baseline Requirements Hinder Trades in Water Quality Trading Programs?" *Journal of Environmental Management* 92(8):2076-2084.

Gollehon, N., M. Caswell, M. Ribaudo, R. Kellogg, C. Lander, and D. Letson. 2001. *Confined Animal Production and Manure Nutrients.* Agricultural Information Bulletin No. 771, U.S. Department of Agriculture, Economic Research Service. (Available online at www.ers.usda.gov/publications/aib771/)

Goolsby, D.A., W.A. Battaglin, B.T. Aulenbach, and R.P. Hooper. 2001. Nitrogen Input to the Gulf of Mexico," *Journal of Environmental Quality* 30: 329-336.

Greene, W.H. 2000. *Econometric Analysis, Fourth Edition.* Upper Saddle River, NJ: Prentice Hall.

Griliches, Zvi. 1958. "The Demand for Fertilizer: An Economic Interpretation of a Technological Change," *Journal of Farm Economics* 40(3): 591-606.

Grossman, M.R. 2007. Agriculture and the Polluter Pays Principle," *Electronic Journal of Comparative Law* 11(3), 1-66.

Hansen, Lars Garn. 2004. "Nitrogen Fertilizer Demand From Danish Crop Farms: Regulatory Implications of Heterogeneity," *Canadian Journal of Agricultural Economics* 52(3): 313-313.

Hefting, M.M., R. Bobbink, and H. de Caluwe. 2003. "Nitrous Oxide Emission and Denitrification in Chronically Nitrate-Loaded Riparian Buffer Zones," *Journal of Environmental Quality* 32:1194-1203.

Henandez, M., and W. Mitsch. 2006. "Influence of Hydrologic Pulses, Fooding Frequency, and Vegetation on Nitrous Oxide Emissions From Created Riparian Marshes," *Wetlands* 26(3): 862-877.

Hernandez-Ramirez, G., S.M. Brouder, D.R. Smith, and G.E. Van Scoyoc. 2009. "Greenhouse Gas Fluxes in an Eastern Corn Belt Soil: Weather, Nitrogen Source, and Rotation," *Journal of Environmental Quality* 38:841-854.

Hey, D.L., L.S. Urban, and J.A. Kostel. 2005. "Nutrient Farming: The Business of Environmental Management," *Ecological Engineering* 24:279–287.

Hoag, D.L. and J.S. Hughes-Popp. 1997. "Theory and Practice of Pollution Credit Trading in Water Quality Management," *Review of Agricultural Economics* 19:252-262.

Horan, R.D., and J.S. Shortle. 2011. "Economic and Ecological Rules for Water Quality Trading," *Journal of the American Water Resources Association* 47(1):58-69.

Hutchinson, G.L., A.R. Mosier, and C.E. Andre. 1982. "Ammonia and Amine Emissions From a Large Cattle Feedlot," *Journal of Environmental Quality* 11:288-293.

Iowa Soybean Association. 2008a. "Managing Nitrogen for Economic and Environmental Performance," *On-Farm Network Update* (July):2-11.

Iowa Soybean Association. 2008b. "Understanding Soil Nitrogen Dynamics," *On-Farm Network Update* (October):2-11.

Jacobs, T.J., and J.W. Gilliam. 1985. "Riparian Losses of Nitrate From Agricultural Drainage Waters," *Journal of Environmental Quality* 14:472-478.

Jansson, M., R. Andersson, H. Berggren, and L. Leonardson. 1994. "Wetlands and Lakes as Nitrogen Traps," *Ambio* 23(6):320-325.

Jarecki, M.K., T.B. Parkin, A.S.K. Chan, T.C. Kaspar, T.B. Moorman, J.W. Singer, B.J. Kerr, J.L. Hatfield, and R. Jones. 2009. "Cover Crop Effects on Nitrous Oxide Emission From a Manure-Treated Mollisol," *Agriculture, Ecosystems, and Environment* 134:29-35.

Just, R. 1975. "Risk Aversion Under Profit Maximization," *American Journal of Agricultural Economics* 57(2): 347-352.

Khakural, B.R., and P.C. Robert. 1993. "Soil Nitrate Leaching Potential Indices: Using a Simulation Model as a Screening System," *Journal of Environmental Quality* 22(4):839-845.

Khosla, R., K. Fleming, J. Delgado, T. Shaver, and D. Westfall. 2002. "Use of Site Specific Management Zones To Improve Nitrogen Management for

Precision Agriculture," *Journal of Soil and Water Conservation* 57:513-518.

King, D.M. 2005. "Crunch Time for Water Quality Trading," *Choices* 20(First Quarter):71-76.

King, D.M., and P.J. Kuch. 2003. "Will Nutrient Credit Trading Ever Work? An Assessment of Supply and Demand Problems and Institutional Obstacles," *Environmental Law Reporter* 33:10352-10368

Kirchmann, H., and L. Bergstrom. 2001. "Do Organic Farming Practices Reduce Nitrate Leaching?" *Communications in Soil Science and Plant Analysis* 32:997-1028.

Lander, C.H., D. Moffitt, and K. Alt. 1998. *Nutrients Available for Livestock Manure Relative to Crop Growth Requirements.* Resource Assessment and Strategic Planning Working Paper 98-1, U.S. Department of Agriculture, Natural Resources Conservation Service, February. (Available online at www.nrcs.usda.gov/technical/NRI/pubs/nlweb.html.) (Accessed September 10, 2009.)

Legg, J.O., and J.J. Meisinger. 1982. "Soil Nitrogen Budgets," in Stevenson, F.J. (ed.), *Nitrogen in Agricultural Soils*, Agronomic Monograph 22, American Society of Agronomy, pp. 503-557.

Lichtenberg, E. 2004. "Cost-Responsiveness of Conservation Practice Adoption: A Revealed Preference Approach," *Journal of Agricultural and Resource Economics* 29(3):420-435.

Lichtenberg, E., and B. V. Lessley. 1992. "Water Quality, Cost-Sharing and Technical Assistance: Perceptions of Maryland Farmers," *Journal of Soil and Water Conservation* 47(3): 260-264.

Liu, X., A. Mosier; A. Halvorson, and F. Zhang. 2006. "The Impact of Nitrogen Placement and Tillage on NO^{-}, N_2O, Ch_4 and CO_2 Fluxes From a Clay Loam Soil," *Plant Soil* 280(1):177-188.

Lowrance, R., L.S. Altier, J.D. Newbold, R.R. Schnabel, P.M. Groffman, J.M. Denver, D.L. Correll, J.W. Gilliam, J.L. Robinson, R.B. Brinsfield, K.W. Staver, W. Lucas, and A.H. Todd. 1997. "Water Quality Functions of Riparian Forest Buffer Systems in Chesapeake Bay Watersheds," *Environmental Management* 21:687-712.

Lowrance, R.R., R.L. Todd, and L.E. Asmussen. 1984. "Nutrient Cycling in an Agricultural Watershed: Streamflow and Artificial Drainage," *Journal of Environmental Quality* 13:27-32.

Malhi, S., and M. Nyborg. 1991. "Recovery of 15N-labelled Urea: Influence of Zero Tillage, and Time and Method of Application," *Fertilizer Research* 28(3):263-269.

Matson, P.A., W.J. Parton, A.G. Power, and M.J. Swift. 1997. Agricultural Intensification and Ecosystem Properties," *Science* 277:504-509.

Mayer, P.M., S.K. Reynolds, Jr., and T.J. Canfield. 2005. *Riparian Buffer Width, Vegetative Cover, and Nitrogen Removal Effectiveness: A Review of Current Science and Regulations*. EPA/600/R-05/118. U.S. Environmental Protection Agency, National Risk Management Research Laboratory.

McSwiney, C.P., and G.P. Robertson. 2005. "Nonlinear Response of N_2O Flux to Incremental Fertilizer Addition in a Continuous Maize (Zea mays L.) Cropping System," *Global Change Biology* 11:1712–1719.

Meisinger, J.J., and G.W. Randall. 1991. "Estimating N Budgets for Soil-Crop Systems," in R.F. Follett et al. (ed.), *Managing Nitrogen for Groundwater Quality and Farm Productivity*, American Society of Agronomy, p. 85-124.

Meisinger, J.J., and J.A. Delgado. 2002. "Principles for Managing Nitrogen Leaching," *Journal of Soil and Water Conservation* 57: 485-498.

Meisenger, J.J., J.S. Schepers, and W.R. Raun. 2008. "Crop Nitrogen Requirement and Fertilization," in J.S. Schepers and W.R. Raun (eds.), *Nitrogen in Agricultural Systems*, American Society of Agronomy, Crop Science Society of America, Soil Science Society of America.

Millar, N., G.P. Robertson, P.R. Grace, R.J. Gehl, and J.P. Hoben. 2010. "Nitrogen Fertilizer Management for Nitrous Oxide (N_2O) Mitigation in Intensive Corn (Maize) Production: An Emissions Reduction Protocol for US Midwest Agriculture," *Mitigation and Adaptation Strategies for Global Change* 15:185-204.

Mitsch, W.J., and J.W. Day. 2006. "Restoration of Wetlands in the Mississippi-Ohio-Missouri (MOM) River Basin: Experience and Needed Research," *Ecological Engineering* 26:55-69.

Mitsch, W.J., J.W. Day, Jr., J.W. Gilliam, P.M. Groffman, D.L. Hey, G.W. Randall, and N. Wang. 1999. *Reducing Nutrient Loads, Especially Nitrate-Nitrogen, to Surface Water, Ground Water, and the Gulf of Mexico: Topic 5 Report for the Integrated Assessment on Hypoxia in the Gulf of Mexico*. Decision Analysis Series No. 19, U.S. Department of Commerce, National Oceanic and Atmospheric Administration.

Mosier, A.R. and L. Klemedtsson. 1994. "Measuring Denitrification in the Field," in R.W. Weaver, S. Angle, P. Bottomley, D. Bezdicek, S. Smith, A. Tabatabai, and A. Wollum (eds.), *Methods of Soil Analysis: Part 2, Microbiological and Biochemical Properties*, Soil Science Society of America, pp. 1047-1065.

Mosier A.R., J.W. Doran, and J.R. Freney. 2002. "Managing Soil Denitrification," *Journal of Soil and Water Conservation* 57:505-513.

Musser, W.N., J.S. Shortle, K. Kreahling, B. Roach, W. Huang, D.B. Beegle, and R.H. Fox. 1995. "An Eonomic Analysis of the Pre-Sidedress Nitrogen Test for Pennsylvania Corn Production," *Review of Agricultural Economics* 17(1):25-35.

Nanda, V.P. 2006. "Agriculture and the Polluter Pays Principle," *American Journal of Comparative Law* 54:317-339.

Pennsylvania Department of Environmental Protection, 2008, *Final Trading of Nutrient and Sediment Reduction Credits – Policy and Guidelines, Appendix A,* Doc. No. 392-0900-001, Harrisburg, PA.

Peoples, M.B., J.R. Freney, and A.R. Mosier. 1995. "Minimizing Gaseous Losses of Nitrogen," in P.E. Bacon (ed.), *Nitrogen Fertilization in the Environment*, New York: Marcel Dekker, Inc., pp. 565-602.

Pionke, H.B., and R.R. Lowrance. 1991. "Fate of Nitrate in Subsurface Drainage Waters," in R.F. Follett, D.R. Keeney and R.M. Cruse (eds.), *Managing Nitrogen for Ground Water Quality and Farm Profitability*, Soil Science Society of America, pp. 237-57.

Power, J.F., R. Wiese, and D. Flowerday. 2001. "Managing Farming Systems for Nitrate Control: A Research Review From Management Systems Evaluation Areas," *Journal of Environmental Quality* 30 (November-December):1866-1880.

Rabalais, N.N., R.E. Turner, and W.J. Wiseman, Jr. 2002a. "Gulf of Mexico Hypoxia, A.K.A. "The Dead Zone," *Annual Review of Ecology, Evolution, and Systematics* 33:235-63.

Rabalais, N.N., R.E. Turner, and D. Scavia. 2002b. "Beyond Science Into Policy: Gulf of Mexico Hypoxia and the Mississippi River," *Bioscience* 52(2):129-142.

Rajsic, P., A. Weersink, and M. Gandorfer. 2009. "Risk and Nitrogen Application Levels," *Canadian Journal of Agricultural Economics* 57:223-239

Randall, G.W., J.A. Delgado, and J.S. Schepers. 2008. "Nitrogen Management To Protect Water Resources," in J.S. Schepers, W.R. Raun, R.F. Follett, R.H. Fox, and G.W. Randall (eds.), *Nitrogen in Agricultural Systems.* Agronomy Monograph 49, Soil Science Society of America, pp. 911-946.

Randall, G.W., M.J. Goss, and N.R. Fausey. 2010. "Nitrogen and Drainage Management To Reduce Nitrate Losses to Subsurface Drainage," in J.A. Delgado, and R.F. Follett (eds.), *Advances in Nitrogen Management for Water Quality*, Soil and Water Conservation Society.

Raun, W.R., and J.S. Schepers. 2008. "Nitrogen Management for Improved Use Efficiency," in J.S. Schepers, W.R. Raun, R.F. Follett, R.H. Fox, and G.W. Randall (eds.) *Nitrogen in Agricultural Systems,* Agronomy Monograph 49, Soil Science Society of America.

Ray, S.C. 1982. "A Translog Cost Function Analysis of U.S. Agriculture, 1939-77," *American Journal of Agricultural Economics* 64(3): 490-498.

Ribaudo, M.O. 2009. "Non-Point Pollution Regulation Approaches in the U.S," in J. Albiac and A. Dinar (eds.), *The Management of Water Quality and Irrigation Technologies*, London: Earthscan.

Ribaudo, M., and C. Nickerson. 2009. "Agriculture and Water Quality Trading: Exploring the Possibilities," *Journal of Soil and Water Conservation* 64(1):1-7.

Ribaudo, M.O. and J. Gottlieb. 2011. "Point-Nonpoint Trading – Can It Work?" *Journal of the American Water Resources Association* 47(1):5-14.

Ribaudo, M.O., and R.D. Horan. 1999. "The Role of Education in Nonpoint Source Pollution Control Policy," *Review of Agricultural Economics* 21(2):331-343.

Ribaudo, M., L. Hansen, D. Hellerstein, and C. Greene. 2008. *The Use of Markets To Increase Private Investment in Environmental Stewardship.* Economic Research Report No. 64, U.S. Department of Agriculture, Economic Research Service. (Available online at www.ers.usda.gov/ publications/err64)

Ribaudo, M., N. Gollehon, M. Aillery, J. Kaplan, R. Johansson, J. Agapoff, L. Christensen, V. Breneman, and M. Peters. 2003. *Manure Management for Water Quality: Costs to Animal Feeding Operations of Applying Manure Nutrients to Land,* Agricultural Economic Report No. 824. U.S. Department of Agriculture, Economic Research Service. (Available online at www.ers.usda.gov/publications/aer824)

Ribaudo, M.O., R.D. Horan, and M.E. Smith. 1999. *Economics of Water Quality Protection From Nonpoint Sources: Theory and Practice.* Agricultural Economic Report No. 782, U.S. Department of Agriculture, Economic Research Service. (Available online at www.ers.usda.gov/ publications/aer782)

Roberts, R.K., and E.O. Heady. 1982. "Fertilizer Demand Functions for Specific Nutrients Applied to Three Major U.S. Crops," *Western Journal of Agricultural Economics* 7(2): 265-278.

Robertson, D.M., G.E. Schwarz, D.A. Saad, and R.B. Alexander. 2009. "Incorporating Uncertainty Into the Ranking of SPARROW Model

Nutrient Yields From Mississippi/Atchafalaya River Basin Watersheds," *Journal of the American Water Resources Association* 45(2):535-549.

Rochette, P., D.A. Angers, M.H. Chantigny, N. Bertrand, and D. Cote. 2004. "Carbon Dioxide and Nitrous Oxide Emissions Following Fall and Spring Applications of Pig Slurry to an Agricultural Soil," *Soil Science Society of America Journal* 68:1410-1420.

Roosen, J., and D.A. Hennessy. 2003. "Tests for the Role of Risk Aversion on Input Use," *American Journal of Agricultural Economics* 85(1):30-43.

Schepers, J.S., K. D. Frank, and C. Bourg. 1986. "Effect of Yield Goal and Residual Soil Nitrogen Considerations on Nitrogen Fertilizer Recommendations for Irrigated Maize in Nebraska," *Journal of Fertilizer Issues* 3:133-139.

Schlegel A.J., K.C. Dhuyvetter, and J.L. Havlin. 1996. "Economic and Environmental Impacts of Long-Term Nitrogen and Phosphorus Fertilization," *Journal of Production Agriculture* (9):114-118.

Shaffer, M.J., and J.A. Delgado. 2001. "Field Techniques for Modeling Nitrogen Management," in Follett et al. (ed.), *Nitrogen in the Environment: Sources, Problems, and Management*. Elsevier Science, pp. 391-411.

Shaffer, M.J., J.A. Delgado, C. Gross, R.F. Follett, and P. Gagliardi. 2010. "Simulation Processes for the Nitrogen Loss and Environmental Assessment Package (NLEAP)," in J.A. Delgado and R.F. Follett (eds.), *Advances in Nitrogen Management for Water Quality*, Soil and Water Conservation Society.

Sharpe, R.R., and L.A. Harper. 1995. "Soil, Plant, and Atmospheric Conditions as They Relate to Ammonia Volatilization," *Fertilizer Research* 42:149-153.

Sheriff, G. 2005. "Efficient Waste? Why Farmers Over-Apply Nutrients and the Implications for Policy Design," *Review of Agricultural Economics* 27(4):542-557.

Shortle, J.S. 1990. "Allocative Efficiency Implications of Water Pollution Abatement Cost Comparisons," *Water Resources Research* 26:793-797.

Shumway, R. 1983. "Supply, Demand, and Technology in a Multiproduct Industry: Texas Field Crops," *American Journal of Agricultural Economics* 65(4): 748-760.

Smith, R.A., G.E. Schwarz, and R.B. Alexander. 1997. "SPARROW Surface Water-Quality Modeling Nutrients in Watersheds of the Conterminous United States: Model Predictions for Total Nitrogen (TN) and Total

Phosphorus (TP)." (Available online at http://water sparrow/wrr97/results.html).

Turner, R.E., and N.N. Rabalais. 2003. "Linking Landscape and Water Quality in the Mississippi River Basin for 200 Years," Vol. 53 No. 6, *BioScience* 563-572.

Turner, R.E., N.N. Rabalais, D. Scavia, and G.F. McIsaac. 2007. "Corn Belt Landscapes and Hypoxia of the Gulf of Mexico," in J.I. Nassauer, M.V. Santelmann, and D. Scavia (eds.), *From the Corn Belt to the Gulf: Societal and Environmental Implication of Alternative Agricultural Futures*, Resources for the Future.

U.S. Department of Agriculture, Economic Research Service (USDA, ERS). 2010. "Direct Government Payments by Program, United States, 1996-2009," Farm Income: Data Files GP9609us. (Available online at www.ers.usda.gov/data/farmincome/finfidmuxls.htm#payments)

U.S. Department of Agriculture, Foreign Agricultural Service (USDA, FSA). 2011. "Corn at a Glance." (Available online at www.fas.usda.gov/htp/CP2011/Corn-2011-Final.pdf)

U.S. Department of Agriculture, National Agricultural Statistics Service (USDA, NASS). 2008. *Agricultural Statistics 2008*. (Available online at www.nass.usda.gov/Publications/Ag_Statistics/2008/index.asp.) (Accessed February 17, 2010.)

U.S. Department of Agriculture, Natural Resources Conservation Service (USDA, NRCS). 2010. *Assessment of the Effects of Conservation Practices on Cultivated Cropland in the Upper Mississippi River Basin.* (Available online at ftp://ftp-fc.sc.egov.usda.gov/NHQ/nri/ceap/UMRB_final_draft_061410.pdf)

U.S. Department of Agriculture, Natural Resources Conservation Service (USDA, NRCS). 2006. "Natural Resources Conservation Service Conservation Practice Standard: Nutrient Management," *National Handbook of Conservation Practice.*

U.S. Department of Agriculture – U.S. Environmental Protection Agency. (USDA-EPA) 1999. Unified National Strategy for Animal Feeding Operations. (Available online at http://cfpub.epa.gov/npdes/afo/ustrategy.cfm)

U.S. Environmental Protection Agency (EPA). 1992. *Another Look: National Survey of Pesticides in Drinking Water Wells, Phase II Report*, EPA 579/09-91-020.

U.S. Environmental Protection Agency (EPA). 2009. *National Summary of Impaired Waters and TMDL Information.* (Available online at http:// iaspub.epa.gov/waters10/attains_nation_cy.control?p_report_type=T)

U.S. Environmental Protection Agency (EPA). 2010a. *Inventory of U.S. Greenhouse Gas Emissions and Sinks: 1990-2008*, EPA 430-R-10-006.

U.S. Environmental Protection Agency, Office of Atmospheric Programs (EPA). 2010b. *Methane and Nitrous Oxide Emissions from Natural Sources,* EPA 430-R-10-001. April.

U.S. Environmental Protection Agency, Office of Wastewater Management (EPA). 2007. *Water Quality Trading Toolkit for Permit Writers*, EPA 883-R-07-004, August.

U.S. Environmental Protection Agency, Office of Water (EPA). 2004. *Water Quality Trading Assessment Handbook: Can Water Quality Trading Advance Your Watershed's Goals?* EPA 841-B-04-001.

Vanotti, M.B., and L.G. Bundy. 1994. "An Alternative Rationale for Corn Nitrogen Fertilizer Recommendations," *Journal of Production Agriculture* (7):243-249.

Vitosh, M.L., J.W. Johnson, and D.B. Mengel. 1995. *Tri-State Fertilizer Recommendations for Corn, Soybeans, Wheat and Alfalfa.* Extension Bulletin E-2567, Michigan State University, The Ohio State University, and Purdue University. (Available online at http://ohioline.osu.edu/ e2567/ index.html.) (Verified June 3, 2010.)

Vitousek, P.M., J.D. Aber, R.W. Howarth, G.E. Likens, P.A. Matson, D.W. Schindler, W.H. Schlesinger, and D.G. Tilman. 1997. "Human Alteration of the Global Nitrogen Cycle: Sources and Consequences," *Ecological Applications* 73(3):737-750.

Weaver, R.D. 1983. "Multiple Input, Multiple Output Production Choices and Technology in the U.S. Wheat Region," *American Journal of Agricultural Economics* 65(1): 45-56.

Westermann, D.T., and G.E. Kleinkopf. 1985. "Nitrogen Requirements of Potatoes," *Agronomy Journal* 77:616-621.

Westermann, D.T., G.E. Kleinkopf, and L.K. Porter. 1988. "Nitrogen Fertilizer Efficiencies on Potatoes," *American Potato Journal* 65:377-386.

Williams, J.R., and D.E. Kissel. 1991. "Water Percolation: An Indicator of Nitrogen-Leaching Potential," in R.F. Follet, D.R. Keeney, and R.M. Cruse (eds.), *Managing Nitrogen for Groundwater Quality and Farm Profitability*, Soil Science Society of America, pp. 59-83.

Wisconsin Department of Natural Resources, 2002. *Agricultural Nonpoint Performance Standards and Prohibitions, NR 151, Subchapter 2.*

Woodward, R.T., and R. A. Kaiser. 2002. "Market Structures for U.S. Water Quality Trading," *Review of Agricultural Economics* 24:366-383.

Wu, J., and B.A. Babcock. 1998. "The Choice of Tillage, Rotation, and Soil Testing: Economics and Environmental Implications," *American Journal of Agricultural Economics* 80(3): 494-511.

Wulf, S., M. Maeting, and J. Clemens. 2002. "Application Technique and Slurry Co-Fermentation Effects on Ammonia, Nitrous Oxide, and Methane Emissions After Spreading: II Greenhouse Gas Emissions," *Journal of Environmental Quality* 31:1795-1801.

Xu, C., M.J. Shaffer, and M. Al-kaisi. 1998." Simulating the Impact of Management Practices on Nitrous Oxide Emissions," *Soil Science Society of America Journal* 62:736-742.

Appendix 1. Estimating Water Treatment Costs

We estimated a treatment cost model with data from the 1996 American Water Works Association (AWWA) survey of its members. There are only 52 usable observations for which utilities provided all required data. This is the last survey in which data on costs and water quality (both raw water coming into and finished water going out of the utility) were gathered at the same time by AWWA. We assume this sample is representative of all water treatment plants. The model is a variable cost function with two outputs (one desirable (water) and one undesirable (nitrogen)); four inputs (three variable and one fixed); and nine factors hypothesized to influence production of drinking water (app. table 1.1). The bootstrap method employed uses network density as the stratum—the result of this stratification .is a more homogeneous sample and hence a smaller standard error.

Econometric Specifi Cation of Simple Production Model and Discussion

Bootstrapped z in parenthesis. Significance level of 0.01 and 0.05 denoted by *** and **, respectively. The estimated variable cost function meets most of the theoretical regularity conditions (i.e., it is monotonically increasing in desirable output as well as in variable inputs). The only case in which the desirable theoretical properties of inputs are not met is in the case of capital, which, in variable cost function setting, should be negative.

$$\ln\left(\frac{V}{w_3}\right) = \hat{\beta}_o + \hat{\alpha}_1 \ln y + \hat{\alpha}_2 \ln N + \hat{\beta}_1 \ln\left(\frac{w_1}{w_3}\right) + \hat{\beta}_2 \ln\left(\frac{w_2}{w_3}\right) + \hat{\eta} \ln K$$

2.01***	0.80***	0.03***	0.62***	0.43***	0.03**
(4.55)	(81.50)	(2.69)	(16.82)	(5.22)	2.27)

$$\hat{\delta}_1 netd + \hat{\delta}_2 public + \hat{\delta}_3 dww + \hat{\delta}_4 syssize + \hat{\delta}_5 loc2 + \hat{\delta}_6 loc3 + \hat{\delta}_7 loc4 + \varepsilon \quad (1)$$

-0.00004***	0.14***	0.08***	0.05***	0.21***	0.22***	0.15***
(-94.50)	(4.65)	(4.96)	(4.19)	(5.53)	(4.77)	(2.93)

Appendix Table 1.1. Summary statistics and definitions

Definition (unit) variable	Mean (Variance)	Definition (unit) variable	Mean (Variance)
Variable cost (in \$) VC	8,479,039 (13,477,167)	System type (1 = Distribution and waste water, 0 = Otherwise) *dww*	0.54 (0.50)
Annual water production (in millions of gallons) y	14,449 (23,498)	System size 1 = if population served greater than 100,000, 0 = Otherwise) *syssize*	0.48 (0.50)
Annual salary (in \$) w_1	\$34,353 (\$11,538)	Consumer structure (ratio of residential to total water delivered) *cs*	0.57 (0.23)
Nitrogen abatement (in difference of raw-finished nitrates in water) (in mg/L) N	0.98 (4.04)	Water system location (New England) *loc1*	0.19 (0.40)
Electricity price (in \$ per kilowatt hour) w_2	\$0.05 (\$0.01)	Water system location (Northeast) *loc2*	0.21 (0.41)
Chemicals price in \$ per pound) w_3	0.2 (0.0)	Water system location (South) *loc3*	0.15 (0.36)
Capital (residual rate of return) K	\$ 145,916,037 (217,806,925)	Water system location (Mid-west) *loc4*	0.21 (0.41)
Network density (population served/length of distribution main) netd	1176 (5608)	Water system location (West) *loc5*	0.23 (0.43)
Organizational type (1 = public, 0 = otherwise) public	0.87 (0.34)		

Source: USDA, Economic Research Service using data from 1996 American Water Works Association survey.

The explanation resides in overcapitalization of water utilities—a phenomenon widely observed for regulated utility firms of all kinds. Homogeneity in the cost function is imposed by dividing both input prices and variable costs by price of chemicals. Consistent with the literature on undesirable outputs, the presence of an undesirable byproduct in a production process, in this case nitrogen, implies a higher cost to the utility which it then abates either to meet regulation[7] or more generally to reduce risk to customers.

As to the exogenous effects, network density has a negative effect on variable costs as expected. Also, larger systems have higher variable costs. Public utilities have higher variable costs than investor-owned utilities. This makes sense from the perspective that public firms may have agency and control problems relative to investor-owned enterprises. Operations that have only a distribution function have lower variable costs than those that have both waste water and distribution. All locations have higher variable costs relative to New England.

Derivation of Shadow Cost of Nitrogen Abatement and Discussion

$$\hat{V} = \exp\begin{pmatrix} \hat{\beta}_o + \hat{\alpha}_1 \ln y + \hat{\alpha}_2 \ln N + \hat{\beta}_1 \ln\left(\frac{w_1}{w_3}\right) + \hat{\beta}_2 \ln\left(\frac{w_2}{w_3}\right) + \hat{\delta}_1 netd \\ + \hat{\delta}_2 public + \hat{\delta}_3 dww + \hat{\delta}_4 syssize + \hat{\delta}_5 loc2 + \hat{\delta}_6 loc3 + \hat{\delta}_7 loc4 \end{pmatrix} * w_3 \tag{2}$$

$$SC_N = \frac{\partial \hat{V}}{\partial N} = \hat{V} * \frac{\hat{\alpha}_2}{N} \tag{3}$$

The shadow marginal cost of nitrogen abatement is derived in equation (3) by taking the derivative of (2), estimated variable cost, which in turn was derived by taking the exponential of (1). From equation (3), various additional derivations can be made: shadow marginal cost by millions gallons, $\partial \hat{V} / \partial N \, / \, y$, estimated shadow total variable cost of nitrogen abatement (*SVC*), $\partial \hat{V} / \partial N \times N$, and *SVC* per millions of gallons of water produced $(\partial \hat{V} / \partial N \times N) \, / \, y$.

The results from the above derivations were used to estimate nitrogen removal costs by system size (app. table 1.2).

Appendix Table 1.2. National estimates of nitrogen removal costs for community water systems, by system size

System size (SS) [in millions of gallons per year]	Estimated average production by CWS	Estimated average cost of nitrogen removal	Estimated total cost of nitrogen abatement
(CWS population in parenthesis)	*(millions of gallons per year)*	*(variable cost per million gallons per year per CWS)*	*(million $ per year for all systems)*
SS > 0 and SS <= 3,300 (42,624)	570	$34.2 [46 %]1	830
SS > 3,300 and SS <= 10,000 (4,871)	4,797	$25.55 [41 %]1	597
SS > 10,000 (4,156)	42,485	$19.18 [31 %]1	3,386

CWS = Community Water System.

[1] Percent of cost attributable.

Source: USDA, Economic Research Service using data from 1996 American Water Works Association survey.

Appendix 2. Using NLEAP to Model Nitrogen Losses

The Nitrogen Loss and Environmental Assessment Package (NLEAP) (Delgado et al., 2010a; Shaffer et al., 2010) can be used to assess the potential for management practices to increase nitrogen use efficiency and generate nitrogen savings that can be traded in water and air quality markets (Delgado et al., 2008b; 2010a). The NLEAP model has been used extensively across national and international systems (Delgado et al., 2008b).

This tool is capable of simulating the effects of management practices and generating reasonable assessment values that are similar to measured field studies conducted across small-scale plots and large commercial field operations (e.g., water budgets, nitrate leaching, residual soil nitrate, crop uptake, nitrogen dynamics, and N_2O emissions; Beckie et al., 1995; Khakural and Robert 1993; 2001; Delgado et al., 2001; Xu et al., 1998).

Detailed descriptions of NLEAP-GIS capabilities and limitations can be found in Shaffer and Delgado, 2001; Shaffer et al., 2010; Delgado and Shaffer,

2008; and Delgado et al., 2010a; 2010b. This improved version can quickly evaluate multiple long-term scenarios across a large number of soils and conduct assessments of the effects of BMPs on nitrogen use efficiency and nitrogen losses via different pathways. The new NLEAP-GIS tool also has a Nitrogen Trading Tool option (with GIS capabilities) (Delgado et al., 2008a; 2008b; 2010a; 2010b).

General Assumptions

NLEAP has been tested, calibrated, and used to accurately evaluate the effects of management for cropping systems and risky landscape combinations across national and international agroecosystems. In order to evaluate these systems, users established basic assumptions to simplify the evaluation process, which is very complex due to the nature of the nitrogen cycle and management interactions with environmental factors (Shaffer and Delgado, 2001).

Yields: It is well known that yield variability can impact nitrogen use efficiency (Bock and Hergert, 1991). Instead of using the maximum yields at a given site as traditionally done by farmers as a safety net approach to calculating nitrogen inputs (Bock and Herget, 1991), State average yields for corn and soybeans derived from the USDA Census of Agriculture were used for the NLEAP-GIS simulations.

We assumed that yields for no-till systems were 10 percent lower than those for conventional tillage. Since we also evaluated excessive nitrogen input scenarios and low nitrogen input (deficit) scenarios, we used the corn yield and nitrogen input response curve from Bock and Herget (1991) to estimate the average yields for these scenarios. It was assumed that for the excessive nitrogen input rates, yields were increased by only 1 percent; however, for the deficit nitrogen input scenario a 10-percent drop in average yield was assumed (Bock and Herget, 1991). We believe that our approach of using average yields to evaluate the effects of management on the nitrogen use efficiency of commercial systems is a valid approach, as reported by Shaffer and Delgado (2001), Delgado (2001), and Delgado et al. (2000; 2001).

Since the USDA Census of Agriculture does not report yields by soil type, we assumed that yields for the soil types tested were similar. However, corn yield can vary among soil type, with lower yields in the sandier, less fertile soils that have higher nitrate leaching potential than those finer soils with lower leaching potential (Khosla et al., 2002; Bausch and Delgado, 2003;

Delgado and Bausch, 2005; Delgado et al., 2005). Nonetheless, we still believe that assuming average yields for a 24-year period being evaluated is a valid approach to assessing the trends and effects of management practices on these different soil types and produces results that are in agreement with average measured values (Delgado et al., 2001; 2008b; 2010a). If additional site-specific field information for a given farm is needed, spatial soil maps for the given farm can be downloaded from USDA NRCS websites, and evaluations using farmers' inputs can be conducted.

Nitrogen Inputs and Uptake: For nitrogen rates, we used the recommended best management practices for site-specific State and/or soil as described by Espinoza and Ross (2008) for Arkansas; Alley et al. (2009) for Virginia; Beegle and Durst (2003) for Pennsylvania; and Vitosh et al. (1995) for Ohio. We calculated the recommended nitrogen (N) rate per bushel of corn derived from each State's recommended BMPs (Espinoza and Ross, 2008; Alley et al., 2009; Beegle and Durst, 2003; Vitosh et al., 1995). A summary of the nitrogen inputs simulated is presented in appendix table 2.1.

Since nitrogen fertilizer inputs were calculated based on yield, the no-till systems received lower nitrogen fertilizer inputs than the conventional systems. However, since a similar rate of uptake per unit of bushel was used for both systems, the removal of nitrogen in harvested grain from the no-till system was also lower than the removal of nitrogen in the grain from the higher yield conventional system. Total nitrogen uptake by the plant was calculated. Initial surface residue cover was simulated at 100, 90, 40, and 30 percent for no-till corn-corn, no-till corn-soybeans, conventional corn-corn, and conventional corn-soybeans, respectively.

For the manure system, manure was applied every 2 years. For the corn-corn rotation, manure was applied in the first year, and only fertilizer was applied in the second year. The manure rate was calculated for each system to match the fertilizer rate. However, since manures will have a large fraction of organic nitrogen that is not immediately available (Davis et al., 2002; Eghball et al., 2002), an additional 50 percent of the recommended rate was added as inorganic nitrogen fertilizer. In other words, the total nitrogen input during the first year of corn-corn rotation was 150 percent of the total application rate of the inorganic nitrogen fertilizer scenario (app. table 2.1). The corn-corn rotation did not receive any manure application in the second year, and the corn received the same rate of nitrogen fertilizer as in the nitrogenfertilizer-only scenario. Thus, over the 2-year period, the manure scenario for corn-corn received an average of 25 percent more nitrogen input per year. The same relationships apply to the excessive and deficit nitrogen scenarios (app. table 2.1).

Appendix Table 2.1. Relationships used to develop yields and nitrogen (N) rates used across the study sites

Tillage	Best management practice	Excessive	Deficiency
	Yield (bushels per acre)		
Conventional	x[1]	x*1.01	x*0.9
No-till	x*0.9	x*0.9*1.01	x*0.81
	N rate for fertilizer-only scenarios (lbs N per acre)		
Conventional	x[2]	z*1.75	z*0.75
No-till	y[3]	y*1.75	y*0.75
	N rate for manure with N fertilizer scenarios (lbs N per acre)		
Conventional	z(org) + 0.5z(fert)	1.75z(org) + z(0.875)	0.75z(org) + z(0.375)
No-till	y(org) + 0.5y(fert)	1.75y(org) + y(0.875)	0.75z(org) + z(0.375)

[1] The x values were 131, 101, 103, and 107 corn bushels per acre for OH, VA, PA, and AR, respectively. The x values were 40, 27, 37 and 27 soybean bushels per acre for OH, VA, PA, and AR, respectively.

[2] The z values were 132, 121, 100, 120, and 125 lbs of N per acre for OH, VA, PA, AR (Hydrology A) and AR (Hydrology D), respectively, for conventional tillage.

[3] The y values were 116, 109, 90, 100, and 105 lbs of N per acre for OH, VA, PA, AR (Hydrology A) and AR (Hydrology D), respectively, for conventional tillage.

Source: USDA, Economic Research Service using data from USDA, Agricultural Research Service.

For the corn-soybean rotation, there was no application of nitrogen fertilizer or manure for any of the scenarios during the soybean year (app. table 2.1). Additionally, for this rotation, the nitrogen cycling from the leguminous soybean crop was credited, as is recommended for each State, so the calculated nitrogen inputs for the corn in the corn-soybean rotation was lower than in the corn-corn system.

The excessive nitrogen fertilizer scenarios received 75 percent higher nitrogen inputs than the State-recommended rate. For the deficit nitrogen application scenarios, nitrogen inputs were applied at a 25-percent lower nitrogen rate than the best management practice scenario (app. table 2.1).

Soil Type Physical and Chemical Information: For each State, the county's soil chemical and physical information averages for the selected soils were downloaded. To evaluate all of the management scenarios described above, we selected a soil with a higher leaching potential (Hydrology A or B) and a soil with a lower leaching potential (Hydrology C or D).

Long-Term Weather: Long-term USDA, Natural Resources Conservation Service weather databases for each county were used to conduct the 24-year assessment as described by Delgado et al. (2008b, 2010a) nitrogen trading tool evaluations.

Other Best Management Practices Tested: For all the scenarios described above, we evaluated the method of application. The best management practice for method of application was incorporation of nitrogen fertilizer and/ or manure. Surface application without incorporation was found to be a poor management practice. We also evaluated time of application. The best management practice for time of application was application of manure and/ or nitrogen fertilizer before planting, closer to the time of higher demand by the crop. The poor management scenario was application of manure and/or fertilizer the previous fall, when the nitrogen is more susceptible to losses.

Long-Term Evaluations: All these scenarios were evaluated over the long term. To conduct the long-term evaluations, we used a 24-year period using long-term weather data for the given county. Similar to what was done with the nitrogen trading tool, the first 12 years were used to run the model, and years 13 to 24 were used to evaluate the effect of management practices on nitrogen use efficiency and on reactive losses to the environment (Delgado et al., 2008b, 2010a).

Appendix 3. Estimating Changes in Nitrogen Fertilizer Application Rate

This appendix describes the econometric model used to estimate changes in nitrogen (N) fertilizer application rate. We estimate nitrogen application rates using an instrumental variables (IV) approach to overcome identification issues presented by farmer heterogeneity and endogenous soil N-testing. Price plays an important role in the nitrogen management decision, and the recent price growth of nitrogen has implications for nitrogen management behavior and by extension, nitrogen use efficiency (NUE). Notably, we instrument for nitrogen price using a cross-section of data by exploiting exogenous spatial variation between domestic ammonia production plants and cornfield locations.

Research using observational data presents econometric challenges, and this is particularly true for research examining the effect of potentially endogenous variables on a study population. For example, when estimating the effect of N-soil tests on application rate, researchers do not know why two observationally identical farmers make different choices about testing the soil. The underlying problem is the concern that unobserved farmer characteristics are responsible for determining whether the farmer conducts a test. For example, a farmer who tests the soil regularly may also have unobserved preferences for land stewardship. If differences beyond observed field, farm operation, and operator characteristics play a role in determining who conducts the test and how the test is used, then the test may be endogenous to the amount applied.

Nitrogen price also presents a challenge in a sample of microdata. Prices are likely to embody an error-in-variables problem because in the case of ARMS, they were created as a share variable that represents the nitrogen fertilizer's relative size of the total expenditures for all fertilizer (nitrogen, phosphorus, and potassium). To see how this effects the estimation of nitrogen demand, consider that we observe nitrogen price as a function of the true, unobserved price plus a disturbance term, *v*.

$$\text{Price}_{\text{N}}\text{Observed} = \text{Price}_{\text{N}}\ \text{True}^{*} + v. \qquad (1)$$

Because the observed price on the left-hand side of equation (1) is a function of true price and *v*, an ordinary least squares (OLS) model of nitrogen demand estimated with the observed price will include *v* and will cause the estimate to be biased and inconsistent. Specifically, in the classic errors-in-variables example, the coefficient in an OLS model will be biased toward zero.[8] Prices farmers pay may also change with their level of demand. For example, if farmers receive quantity discounts when purchasing nitrogen fertilizer and their application rate is correlated with total nitrogen demand, then failing to account for this also results in bias.

To overcome the problem of mismeasured nitrogen prices and endogenous soil testing, we employ an IV approach, which allows for the development of consistent and unbiased estimates. In the case of endogenous N-soil testing, we find a set of instruments that are correlated with N-soil testing but uncorrelated with disturbance process: average annual soil percolation and average annual precipitation. Because percolation facilitates nutrient leaching (Williams and Kissell, 1991), we expect the greater soil percolation to increase uncertainty about available nutrients, and, therefore, encourage soil testing.

Higher precipitation generally reduces the ability to conduct soil test, therefore we expect annual average precipitation to be negatively related to N-soil test.

We identify the nitrogen own-price effect on demand using three sources of exogenous variation: distance between the field and domestic ammonia fertilizer production; production capacity of nearby ammonia plants; and distance from the field to New Orleans, LA, site of the majority of international ammonia importation.[9] Ammonia is increasingly being imported by the United States, and a majority of shipments enter from the Gulf of Mexico, and specifically, New Orleans; therefore, we also include a distance-to-New Orleans measure. These variables are useful instruments because the distance between the field and production capacity are arguably uncorrelated with the behavior of the farmer or the placement of the field;[10] therefore, the instruments allow one to capture the exogenous variation in price and use it to estimate application rates.

Instrumental Variables Model

We use an IV model specified with two endogenous variables to estimate a partial-equilibrium static demand model derived from profit maximization theory. The model assumes producers make immediate adjustments to quantity demanded in response to changes in price, and that prices are known at the time of production planning. These assumptions are reasonable given the ability of farmers to enter into contracts that establish price for delivered corn and inputs to production, such as forward or marketing contracts, and other hedging instruments. Further, production technology is assumed known and fixed. Since only two time periods separated by 4 years are used, technology is unlikely to change. The most likely technological change is that of seed technology—the use of biotech (Bt) corn; however, the model specification controls for this. In 2001, 20 percent of corn acres were planted with Bt corn; in 2005, the amount was slightly greater than 30 percent.

We characterize the problem posed to the farmer as one of profit maximization with uncertainty, as evidenced by the nitrogen overtreatment, but the decision of the farmer could also be conceptualized as a utility maximization problem. In this case, the farmer chooses a level of output that maximizes the farmer's initial wealth plus expected profit from the operation. Under utility maximization, a farmer considers not only expected profit but moments of the profit distribution as well, and deviations from the recommended level of nitrogen then depend on the farmer's level of risk aversion. Evidence from field trial suggests that risk-neutral farmers would be willing to overapply nitrogen to increase profits during a year of "good" growing conditions (Rajsic et al., 2009). On the other hand, risk-averse farmers will reduce their nitrogen rate to reduce profit variance. In practice, our empirical results are not dependent on the conceptual framework; in both cases, nitrogen prices enter the profit function, and the identification strategy would not change. Rather, the level of risk aversion primarily drives the differences. Some research, however, suggests that risk-averse farmers are more responsive to price because of profit risk (Just, 1975; Roosen and Hennessy 2003; Rajsic et al., 2009), and, if farmers are on average risk averse, our elasticity estimates will represent an upper bound.

Equation (2) is the outcome equation where V represents the log transformed per acre rate of nitrogen applied to the field of farm i in USDA production region r at time t. Endogenous variables, $T^{\hat{}}$ and $P^{\hat{}}$, are estimated N-soil testing probability and nitrogen price from equations (3) and (4). The set of excluded instruments for N-soil test are represented by Z^T, and the excluded instruments used to estimate nitrogen price are represented by Z^P. The vector $\boldsymbol{X}$ is a set of independent variables that includes characteristics of the operator, farm operation, and the field; the disturbance term is represented by ε.

$$Y_{irt} = \alpha_1 + \hat{T}_{irt}\beta_1 + \hat{P}_{irt}\lambda_1 + \mathbf{X}_{irt}\delta_1 + \phi_{1r} + \upsilon_{1t} + \varepsilon_{irt}\,,$$

$$T_{irt} = \alpha_2 + \mathbf{X}_{irt}\beta_2 + Z^T{}_{irt}\delta_2 + \phi_{2r} + \upsilon_{2t} + \kappa_{irt}\,,$$

$$P_{irt} = \alpha_3 + \mathbf{X}_{irt}\beta_3 + Z^P{}_{irt}\delta_3 + \phi_{3r} + \upsilon_{3t} + u_{irt}\,.$$

A case can be made that countrywide trends over time affect the use of nitrogen. Perhaps in response to outreach efforts to reduce fertilizer runoff due to overuse, for example, environmental awareness campaigns that communicate the benefits of reduced nitrogen in the environment, attitudes about nitrogen rates have changed. We control for trends in nitrogen use that change over time with a time effect term, υ_t.

As well, use of nitrogen across production USDA-defined regions may also affect application rates, therefore we control for region-specific factors with a fixed-effect term, ϕ_r.

Data

The data are cross-sectional and come from USDA's Agricultural Resource Management Survey (ARMS). ARMS comprises responses to a series of interviews with farm operators designed to solicit information about production practices, costs of production, business finances, and operator and household characteristics. Commodity specific surveys are fielded on a rotating basis, usually every 5 to 8 years. We focus on corn production because of its intense use of nitrogen, for which ARMS last fielded surveys in 2001 and 2005.

We use data from two components of ARMS. The first component is the *Corn Production Practices and Costs Report*, which surveys the farm enterprise's costs of production and a host of production practices at the field level. The second component is the *Corn Costs and Returns Report*, which collects indepth financial information concerning the farm business and the household of the operator. The two components can be linked together to provide a complete view of the farm operation from the farm's representative field to its financial statement, and we restrict the sample to farmers who completed both surveys.

As covariates, we include the farmer's age, education, and income earned from work off the farm. We account for land quality and tenancy issues by including the per acre annual value of production, the per acre value of the land, and acres owned by the operator. We also control for environmental characteristics of the field, for example, whether any part of the field is a classified as a wetland. The presence of livestock and a nutrient management plan on the farm may indicate a greater reliance on manure, driven often by the need to dispose of manure.

We account for these with dummy variables as well. The nutrient requirements of a current corn crop are also based, in part, on the plant-available nutrients existing in the soil, and past cropping practice can influence these nutrients. Therefore, we use a dummy variable to control for crop rotation pattern of 3-year straight corn rotation.

The timing and method of application may also be important determinants of application rate. A spring application is better timed to meet the plant's need for nutrients and reduces the risk of loss due to environmental factors relative to a fall or winter application.

On the other hand, farmers may opt to apply nitrogen in the fall, when there are fewer time demands and prices are often lower. In such a case, a nitrogen inhibitor is often used to further slow the nitrification process, though average annual nitrate losses can still be 50 percent higher under fall application than under spring application (Randall and Mulla, 2001).

To counter this, in many cases, anhydrous ammonia is injected into the soil because low temperatures at this time of year slow the conversion of ammonia to ammonium and nitrate, reducing the loss of nitrogen. We control for the method of application with a dummy variable indicating whether the nutrient was incorporated or injected into the soil.

Technology and other management practices thought to affect nitrogen rate are captured by explanatory variables indicating the use of field irrigation and biotech (Bt) corn seed. Irrigation is an important component in nitrogen management. Irrigation may be a necessary practice due to the climate, or it may be another way of more precisely controlling growing conditions. If water and nitrogen are complementary inputs, the presence of irrigation should increase the rate of nitrogen application.

The use of biotech seed is driven by the associated cost reductions from the technology's herbicide, pest, or fungus resistance. We also include a dummy variable representing whether the corn crop was grown for silage or corn. A full list of covariates and summary statistics is presented in appendix table 3.1.

Outcome Measures

We estimate the application rate for four different permutations of nitrogen fertilizer use. First, we estimate commercial nitrogen use by farmers who exclusively apply commercial nitrogen—a group that accounts for a 78 percent of the farmers in our sample. We also examine the rate of total commercial nitrogen use by all farmers, regardless of whether they used commercial nitrogen exclusively or in conjunction with manure.

The third measure examines the sensitivity of commercial nitrogen use by farmers who use manure in conjunction with commercial nitrogen—a group that employs an imperfect substitute for commercial nitrogen. These farmers make up a minority of the sample, 22 percent.

Finally, we examine the effect of our explanatory variables on total nitrogen application rate, which includes commercial nitrogen and manure. It should be noted that all of the farmers in the sample reported at least some use of commercial nitrogen fertilizer. Estimates from the IV model are presented in appendix table 3.2.

Appendix Table 3.1. Summary statistics

Variable name	Description	Mean	95% confidence	Interval
Soiltestn	Nitrogen soil test	0.21	0.18	0.24
Nprice	Nitrogen price	0.328	0.324	.332
Dealerrec	Dealer recommendation	0.32	0.29	0.35
Consultrec	Consultant recommendation	0.14	0.12	0.16
Extrec	Extension agent recommendation	0.04	0.02	0.05
Routine	Routine practice	0.28	0.26	0.30
op_age	Operator's age	52.73	52.11	53.36
Retired	Operator is retired from farming	0.04	0.03	0.06
College	Operator holds college degree	0.35	0.31	0.37
Workoff	Derive income from off-farm work	0.38	0.35	0.42
Anycropins	Insurance participation rate	0.659	0.62	0.70
Prodvalpa	Production value per acre	$4,	372.57	$337.29
Landvalpa	Land value per acre	$1,616.55	$709.46	$2,523.64
Ownacre	Acres owned	323.37	301.10	345.63
Corn_p	Corn price	1.87	1.84	1.90
CCC	Straight corn rotation (3 years)	0.25	0.21	0.28
Nutrient plan	Nutrient plan in place	0.076	0.063	0.088
Irrigate	Irrigate the field	0.063	0.0397	0.0853
Wetland	Wetland on any part of the field	0.03	0.02	0.04
Tenure	Years farming	27.61	26.89	28.33
Spring	Spring fertilizer application	0.80	0.77	0.84
Inc	Incorporated fertilizer	0.75	0.73	0.78
Inhibit	Fertilizer applied with inhibitor	0.07	0.05	0.09
Bt_corn	Biotech corn	0.34	0.30	0.38
Yldgoal	Yield goal	173.62	166.31	180.94
Silage	Corn for silage	0.11	0.09	0.13
Livestock	Presence of livestock on the farm	0.576	0.55	0.602
Commercial nitrogen w/o manure	Commercial nitrogen users only	129.72	125.67	133.77
Total commercial nitrogen	Total commercial nitrogen use	118.42	114.42	122.42
Commercial nitrogen w/ manure	Commercial nitrogen use by manure users	77.23	70.60	83.87
	Total commercial nitrogen and manure use	137.59	132.16	143.02
Total nitrogen observations		2,874		

Source: USDA, Economic Research Service using data from USDA's 2001 and 2005 Agricultural Resource Management Survey.

Appendix Table 3.2. IV estimates of nitrogen application rate

	Commercial nitrogen: nonmanure users	S.E.	Total commercial nitrogen	S.E.	Commercial nitrogen: only manure users	S.E.	Total nitrogen (manure and nonmanure users)	S.E.
Soiltestn	-0.924**	0.290	-1.142**	0.336	0.333	0.742	-1.080**	0.308
Lognprice	-1.347	0.715	-1.379*	0.630	0.531	1.408	-0.674	0.589
Dealerrec	0.131**	0.043	0.159**	0.047	0.155	0.099	0.157**	0.042
Consultrec	0.229**	0.078	0.291**	0.083	-0.004	0.171	0.303**	0.078
Extrec	0.084	0.086	0.143	0.084	0.239	0.156	0.163*	0.073
Routine	-0.170**	0.065	-0.164**	0.063	-0.071	0.100	-0.136**	0.057
Op_age	-0.011**	0.003	-0.008**	0.003	0.005	0.007	-0.008**	0.002
Retired	0.104	0.098	0.107	0.100	0.171	0.211	0.002	0.089
College	0.043	0.034	0.055	0.037	0.118	0.117	0.025	0.034
Workoff	-0.091**	0.037	-0.0810*	0.0398	-0.124	0.094	-0.115**	0.037
Anycropins	0.061	0.054	0.1065*	0.0498	0.035	0.083	0.1203**	0.0468
Prodvalpa	-7.84E-06	3.32E-05	-3.09E-05	2.50E-05	-4.38E-05	3.25E-05	4.64E-05**	1.98E-05
Landvalpa	-4.93E-07	4.41E-07	-8.96E-07	7.65E-07	-5.21E-05	2.81E-05	-1.57E-06	8.10E-07
Ownacre	3.62E-05**	1.37E-05	3.92E-05**	1.48E-05	-6.32E-05	5.88E-05	2.95E-05	1.51E-05
logcorn_p	0.006	0.043	0.029	0.048	-0.032	0.112	0.034	0.045
Ccc	0.0315	0.055	0.092	0.052	0.192**	0.088	0.082	0.051
Wetland	-0.081	0.118	-0.065	0.110	-0.187	0.326	-0.013	0.098
Nutrplan	0.167**	0.070	0.023	0.077	-0.280**	0.133	0.172**	0.072
Irrigate	0.527**	0.085	0.532**	0.089	-0.370	0.364	0.551**	0.084
Tenure	0.006**	0.002	0.005**	0.002	0.005	0.007	0.004	0.002
Spring	0.028**	0.041	0.013	0.048	-0.083	0.150	0.026	0.042
Inc	0.063	0.050	0.061	0.049	0.052	0.101	0.053	0.046
Inhibit	0.083	0.057	0.2239**	0.0590	0.556**	0.120	0.176**	0.056
Bt_corn	0.042	0.036	0.067	0.040	0.081	0.100	0.062	0.036
Yldgoal	0.001**	0.0002	0.0003	0.0002	-5.27E-06	2.23E-04	0.0001	0.0002

Appendix Table 3.2. (Continued)

	Commercial nitrogen: nonmanure users	S.E.	Total commercial nitrogen	S.E.	Commercial nitrogen: only manure users	S.E.	Total nitrogen (manure and nonmanure users)	S.E.
Silage	-0.404**	0.093	-0.350**	0.078	-0.060	0.094	-0.098	0.076
Live	-0.154**	0.046	-0.233**	0.051	-0.265	0.186	-0.142**	0.047
Observations	2253		2874		624		2874	
F-Statistic	6.69		11.87		6.31		7.82	
	[<0.000]		[<0.000]		[<0.000]		[<0.000]	

Source: USDA, Economic Research Service using data from USDA's 2001 and 2005 Agricultural Resource Management Survey.

APPENDIX 4. COMPARING COSTS OF FARMS USING DIFFERENT NUTRIENT MANAGEMENT PRACTICES

The goal of this analysis was to estimate the variable production costs for farms using different nutrient management strategies. The results are used to estimate the cost of changing from a less-efficient to a more-efficient nutrient management strategy. We restricted our analysis to corn, given the large acreage and its intensive use of nitrogen.

Data on corn are from USDA's 2001 Agricultural Resource Management Survey (ARMS). This is the last corn survey from which field-level cost of production data are estimated for each observation. SAS General Linear Model procedure (GLM) was used to estimate a model of variable production costs as a function of management and resource-base variables. Least squares means were used to compare the per acre variable production costs between practices directly related to nitrogen management.

Total variable costs (TVC) were defined as the costs of seed, fertilizer, manure, pesticides, custom work, and fuel lubricants. We specified a model of TVC as a function of the following variables:

1. Use of biotech or herbicide resistant corn
2. Use of rotation with soybeans
3. Use of nitrogen inhibitor
4. Tillage (conventional till vs. reduced/no till)
5. Timing (fall vs. spring application)
6. Method (broadcast vs. inject/incorporate)
7. Conservation cropping (contour or strip)
8. Presence of nutrient management plan
9. Use of variable rate technology
10. Presence of irrigation
11. Presence of highly erodible soils (yes or no)
12. Presence of tile drains
13. Growing season (northern tier, middle tier, southern tier)
14. Farm size (total corn acres on farm)
15. Yield goal

An interaction term for timing and method (fall/no fall – incorporate/ broadcast) was also included. The cost model was run separately for those farms that do not use manure and for those farms that use both manure and commercial fertilizer. About 16 percent of U.S. corn acres receive manure.

Appendix Table 4.1. Variable cost per acre of management practices

	Commercial nitrogen only		Commercial nitrogen and manure	
Management choice	*Dollars per acre*	*Pr > t*	*Dollars per acre*	*Pr > t*
Continuous corn	131.23	.0001	165.63	.1330
Rotation with soybeans	124.02		158.06	
Conventional tillage	128.79	.1554	128.79	.6671
Reduce/no-till	126.46		126.46	
Fall/broadcast	127.84	.0582	158.89	.1587
Fall/incorporate	128.39	.0557	155.25	.1867
Spring/broadcast	132.54	.0001	158.53	.2078
Spring/incorporate	121.74		174.70	
No irrigate	133.33	.0009	164.11	.7292
Irrigate	121.92		159.58	
No highly erodible soil	124.92	.0088	157.98	.2013
Highly erodible soil	130.34		165.71	
No nitrogen inhibitor	125.34	.0832	153.80	.0441
Nitrogen inhibitor	129.92		169.88	
No conservation cropping	131.83	.0004	163.25	.6278
Conservation cropping	123.42		160.44	
No nutrient plan	127.87	.8475	157.63	.1461
Nutrient plan	127.39		166.06	
No variable rate	125.45	.1522	155.39	.2945
technology	129.81		168.30	
Variable rate technology				
No tiles	128.79	.2095	168.02	.0366
Tiles	126.46		155.67	

Source: USDA, ERS using USDA's 2001 Agricultural Resource Management Survey.

Since most of the variables are class variables, we used the SAS General Linear Model procedure (GLM) to estimate the model. The R-Squares of the no-manure and manure-cost models are 0.21 and 0.16, respectively, and the models are significant at the 1-percent level. The majority of the explanatory variables are statistically significant at the 5-percent level. Least-square means of the production costs ($/acre) under the different management systems are presented in app. table 4.1, along with an indication of whether the difference is statistically significant. Of interest to this study is that the cost under the preferred method/timing combination (spring/incorporate) is significantly different from the costs under the less-preferred, alternative combinations (at the 5- and 10-percent levels) for those farms that use only commercial fertilizer (84 percent of treated corn acres). No significant differences in costs were found for farms that use both manure and commercial fertilizer.

Appendix Table 4.2. Nitrogen application rates per acre by management practice

	Commercial nitrogen only		Commercial nitrogen and manure	
Management choice	*Pounds per acre*	*Pr > t*	*Pounds per acre*	*Pr > t*
Continuous corn	136	.1544	218	.0420
Rotation with soybeans	140		192	
Conventional tillage	137	.3583	202	.6433
Reduce/no-till	139		208	
Fall/broadcast	143	.0001	191	.0420
Fall/incorporate	141	.0042	201	.6433
Spring/broadcast	140	.0001	220	.8382
Spring/incorporate	129		208	
No irrigate	129	.0002	210	.7692
Irrigate	147		200	
No highly erodible soil	139	.5955	222	.0353
Highly erodible soil	137		188	
No nitrogen inhibitor	135	.0017	189	.1354
Nitrogen inhibitor	141		221	
No conservation cropping	143	.0004	175	.0001
Conservation cropping	133		235	
No nutrient plan	137	.6002	197	.2950
Nutrient plan	139		213	
No variable rate technology	138	.9899	224	.3175
Variable rate technology	138		186	
No tiles	139	.4957	214	.2384
Tiles	137		196	

Note: Parameter estimates from GLM model.

Source: USDA, Economic Research Service using data from USDA's 2001 Agricultural Resource Management Survey.

Part of the difference in costs observed with ARMS data is due to differences in chemical application rates. Since the NLEAP scenarios assumed the management changes were independent, altering rate, timing, and method in different combinations, we needed to separate out the nitrogen fertilizer cost from the total of changing management. We ran the same models, but with nitrogen application rate as the dependent variable. Both of the models were significant, with R-squares of 0.23 and 0.24. Differences in application rates between the spring/inject and the other management combinations were positive (as expected) and significant at the 1-percent level for farms using only commercial fertilizer (app. table 4.2). The difference in nitrogen fertilizer costs was subtracted from the cost difference derived from the cost model, using a nitrogen fertilizer price of $0.30/lb. The cost of adopting appropriate method (assuming no change in fertilizer application rate) was estimated to be $7.35/acre, appropriate timing was $3.01 per acre, and both appropriate method and timing were $1.86/acre. For farms using manure, we assumed no differences in costs.

APPENDIX 5. ESTIMATING WETLAND RESTORATION COSTS

The cost of restoring a wetland is the sum of the cost of the land and the cost of restoring the land's water-hold capability and the wetland ecosystem. We generate wetland and restoration costs using cost functions that we estimated using available wetland cost data. Sample observations lie in the Glaciated Interior Plains (GIP).

The cost of the land to society is the difference in its value with and without the wetland. The value of agricultural land without a wetland is assumed to be a function of the net value of its output, but the potential for nonagricultural use can play a role.

The USDA Wetland Reserve Program (WRP) sets wetland easement prices equal to the difference in land values with and without a permanent wetland easement. Therefore, WRP easement payments are well suited as a measure of land cost.

Land cost is modeled as a function of the agricultural value and value squared of the land in the contract (AgrValue and AgrValuesq), contract size and size squared (Acres and Acressq), the potential for urban development (Urban), and farm size (Fsize). Because a measure of the agricultural value of the land is not available, we use the product of the county-average farmland rental rate (Rent) and contract acreage as a proxy (it represents the annual agricultural value of the land). The adjusted R-square of the estimated land cost model indicates that the estimated ordinary least squares model explains 90 percent of the variation in WRP land costs. Variables are statistically significant and have the expected sign. With this cost function, we generate marginal and average land cost estimates by county throughout the GIP. To generate average cost, we divide total land cost (generated with our model) by the size of the contract—all cost estimates are based on the median-size WRP easement. Across the counties of the GIP, average per acre costs range from \$1,490 to \$3,030. Second, we generate the marginal cost function (MC_L) by differentiating the estimated land-cost model with respect to Acres:

$$MC_L = 925 + 4.32*Rent + 2.39(10^{-6})*AgrValue*Rent - 0.127*Acres.$$

For average-sized contracts, county-level estimates of MC_L in GIP range from \$985 to \$1,790 per acre with a median cost of \$1,390. Restoration costs are modeled as a function of the agricultural value of the land, the size of the contract, and other variables. The agricultural value is included as an explanatory variable because we believe that landowners would spend more to drain more productive lands and assume that restoration costs are positively correlated with drainage costs. Approximately 15 percent of the WRP contracts of the GIP report zero restoration costs. Because the dependent variable is truncated, we use the Tobit procedure to estimate the restoration cost function. The Tobit procedure simultaneously estimates the probability that the dependent variable is nonzero and its expected size. Variables of the estimated model are statistically significant and have the expected sign.

The estimated model is used to generate expected restoration costs. By dividing our model's county-level expected cost estimates by contract size, we generate estimates of expected average restoration costs. Costs range from \$506 to \$602 per acre across counties. Differentiating the estimated Tobit model with respect to the contract acres generates the expected marginal restoration cost function (MC_R):

$$MC_R = (Z)*(0.888*Rent -2.12* AgrValue*Rent + 167)$$

where (Z) is the cumulative probability function and Z is the estimated Tobit equation. For average-sized contracts, estimates of MC_R across counties of the GIP range from \$101 to \$210 per acre.

End Notes

[1] These four States were selected because they present a wide variation in growing conditions and because the data necessary for running NLEAP were already developed.

[2] Recall that this adoption rate is higher than that reported by the USDANRCS CEAP analysis, which considers adoption over multiyear rotations (see box on page 17).

[3] Because we are comparing 2001 costs with 2008 payments, we inflate 2001 costs using the U.S. Bureau of Labor Statistics' Consumer Price Index.

[4] Based on EPA analysis of the American Clean Energy and Security Act of 2009, H.R. 2454.

[5] The mean commercial nitrogen application rate in our sample was 129.72 lbs per acre.

[6] The ARMS data do not enable us to identify only those program payments subject to compliance, but they are a good approximation.

[7] EPA regulates nitrate in drinking water (measured as nitrogen) at 10 mg/L.

[8] See Greene (2000) for a formal discussion of measurement error and the resulting attenuation bias.

[9] Ammonia production data come from the *North American Fertilizer Capacity Annual Reports* issued by the International Fertilizer Development Center. We calculate the distances from the field to ammonia production using the location of the plant and geocoded corn field samples from USDA's 2001 and 2005 Agricultural Resource Management Survey. It should be noted that these are sample points, and they do not represent all corn production in the United States; however, when we estimate a model of nitrogen demand, the sample points are weighted to reflect total U.S. corn production.

[10] To test that the instruments are uncorrelated with the residual component in the second stage of the IV model, or exogenous to the rate of fertilizer application, we test overidentification restrictions using a Sargan test. The test statistic is computed as $n \times R^2$ and has a $\chi^2(k\text{-r})$ distribution, where k is the number of instruments and r is the number of endogenous variables. The results of the test are presented in the results table.

In: Nitrogen Use in United States Agriculture ISBN: 978-1-62081-636-3
Editors: E.F. Wilkinson, E.J. Parker

Chapter 2

NITROUS OXIDE FROM AGRICULTURAL SOURCES: POTENTIAL ROLE IN GREENHOUSE GAS EMISSION REDUCTION AND OZONE RECOVERY*

Kelsi Bracmort

SUMMARY

Gases other than carbon dioxide accounted for approximately 17% of total U.S. greenhouse gas emissions in 2009, yet there has been minimal discussion of these other greenhouse gases in climate and energy legislative initiatives. Reducing emissions from non-carbon dioxide greenhouse gases, such as nitrous oxide (N_2O), could deliver short-term climate change mitigation results as part of a comprehensive policy approach to combat climate change.

Nitrous oxide is 310 times more potent than carbon dioxide in its ability to affect the climate; and moreover, results of a recent scientific study indicate that nitrous oxide is currently the leading ozone-depleting substance being emitted. Thus, legislation to restrict nitrous oxide emissions could contribute to both climate change protection and ozone recovery.

* This is an edited, reformatted and augmented version of Congressional Research Service, Publication No. R40874, dated December 7, 2011.

The primary human source of nitrous oxide is agricultural soil management, which accounted for more than two-thirds of the N_2O emissions reported in 2009 (approximately 205 million metric tons CO_2 equivalent). One proposed strategy to lower N_2O emissions is more efficient application of synthetic fertilizers. However, further analysis is needed to determine the economic feasibility of this approach as well as techniques to measure and monitor the adoption rate and impact of N_2O emission reduction practices for agricultural soil management.

As the 112th Congress considers legislation that would limit greenhouse gas emissions, among the issues being discussed is how to address emissions of non- CO_2 greenhouse gases. Whether such emissions should be subject to direct regulation, what role EPA should play using its existing Clean Air Act authority, and what role USDA should play in any N_2O reduction scheme are among the issues being discussed. How these issues are resolved will have important implications for agriculture, which has taken a keen interest in climate change legislation.

INTRODUCTION

Policymakers are dedicating considerable attention to greenhouse gas emission reduction, primarily discussing options for carbon dioxide (CO_2) emission reduction.[1] Less frequently addressed in proposed legislation is emission reduction for non- CO_2 greenhouse gases, such as nitrous oxide (N_2O). However, N_2O reduction efforts have the potential to mitigate climate change. Moreover, N_2O emission sources may be regulated under the existing Clean Air Act as a class I or class II ozone-depleting substance at the discretion of the Environmental Protection Agency (EPA) Administrator. No new legislation needs to be passed to regulate N_2O for climate protection and ozone recovery.

The five non- CO_2 greenhouse gases regularly monitored but not entirely regulated by EPA (methane, nitrous oxide, hydroflourocarbons, perflourocarbons, and sulfur hexaflouride) accounted for approximately 17% of U.S. greenhouse gas (GHG) emissions in 2009, as measured by total tons of CO_2 equivalent.[2] Nitrous oxide—the third-most abundant greenhouse gas—was responsible for roughly 4% of total U.S. GHG emissions in 2009 by weight. Although they comprise a smaller portion of GHG emissions, non-CO_2 greenhouse gases, including N_2O, are more potent than CO_2. The gases identified above are 21 to 23,900 times more effective than an equivalent weight of CO_2 at trapping heat in the atmosphere, with N_2O being 310 times more potent by weight.[3]

In addition to being one cause of greenhouse gas emission growth, N_2O is an ozone-depleting substance (ODS).[4] Indeed, scientific analysis suggests that N_2O is now the leading ODS being emitted, as emissions of other substances have been reduced significantly owing to regulations enacted in the late 1980s, in the Montreal Protocol on Substances that Deplete the Ozone Layer.[5] N_2O emission reduction could thus play a compelling role in recovery of the ozone layer as well as in greenhouse gas emission reduction.

The agriculture sector is the primary anthropogenic source of nitrous oxide.[6] The bulk of U.S. N_2O emissions stem from fertilizing agricultural soils for crop production. Strategies or technologies designated for N_2O emission reduction are limited.[7] This is partly due to the dispersed nature of N_2O emission sources.

In the agriculture sector, the majority of N_2O is released as a consequence of specific nitrogen cycle processes (nitrification and denitrification) when large amounts of synthetic nitrogen fertilizers are used for crop production. More efficient application of synthetic fertilizers (e.g., precision agriculture, nitrogen inhibitors, nitrogen sensors, controlled-release fertilizer products) is one way to reduce excess amounts of nitrogen available for bacterial processing and eventual release to the atmosphere as N_2O. High costs and difficulty in measuring these products' efficacy, among other deterrents, have hampered widespread adoption of practices to reduce N_2O emissions.

This report focuses on the contributions of N_2O to alteration in the Earth's climate and ozone depletion. Policy options for N_2O emission reduction, sources of N_2O, and federal support to lower N_2O emissions are discussed.

Nitrous Oxide: A Primer

Nitrous oxide (N_2O), familiar to some as "laughing gas," contributes to climate change and ozone depletion. Once released, N_2O lingers in the atmosphere for decades (its atmospheric lifetime is approximately 114 years) and is 310 times more effective at trapping heat in the atmosphere over a 100-year time frame than carbon dioxide (CO_2).[8] N_2O emission quantity estimates remained fairly constant from 2005 to 2007, hovering around 325 million metric tons carbon dioxide equivalent (CO_2e). N_2O emission quantity estimates dropped in 2009 to below 300 million metric tons CO_2e. See **Table 1**.

Table 1. U.S. Greenhouse Gas Emissions (million metric tons CO_2e)

Gas / Source	2005	2006	2007	2008	2009	Avg. Contribution[a]
Carbon dioxide (CO_2)	6,113.8	6,021.1	6,120.0	5,921.4	5,505.2	83%
Methane (CH_4)	631.4	672.1	664.6	676.7	686.3	9%
Nitrous oxide (N_2O)	322.9	326.4	325.1	310.8	295.6	4.4%
Hydroflourocarbons (HFCs)	120.2	123.5	129.5	129.4	125.7	1.8%
Perflouruocarbons (PFCs)	6.2	6.0	7.5	6.6	5.6	<1%
Sulfur hexaflouride (SF6)	19.0	17.9	16.7	16.1	14.8	<1%
Total	7,213.5	7,167.0	7,263.4	7,061.0	6,633.2	

Source: Environmental Protection Agency, *2011 U.S. Greenhouse Gas Inventory Report*.

[a] Average contribution to total U.S. greenhouse gas inventory based on data provided for 2005 to 2009.

Sources of N_2O Emissions

Nitrous oxide is emitted from anthropogenic (manmade) and natural sources. Oceans and natural vegetation are the major natural sources of N_2O. Agricultural soil management (e.g., fertilization, application of manure to soils, drainage and cultivation of organic soils) is responsible for more than two-thirds of anthropogenic U.S. N_2O emissions.[9]

In 2009, N_2O emissions from agricultural soil management totaled more than 200 million metric tons of CO_2e.[10] Other anthropogenic sources of N_2O are combustion by mobile sources (cars, trucks, etc.), manure management, and nitric acid production.[11]

Figure 1 depicts the origination and passage of nitrogen (N) that leads to N_2O emissions from agricultural soil management. The amount of N_2O emitted from cropland soils largely depends on the amount of nitrogen applied to a crop, weather, and soil conditions.

Corn and soybean crops emit the largest amounts of N_2O, respectively, due to vast planting areas, plentiful synthetic nitrogen fertilizer applications, and, in the case of soybeans, high nitrogen fixation rates (Figure 2).[12]

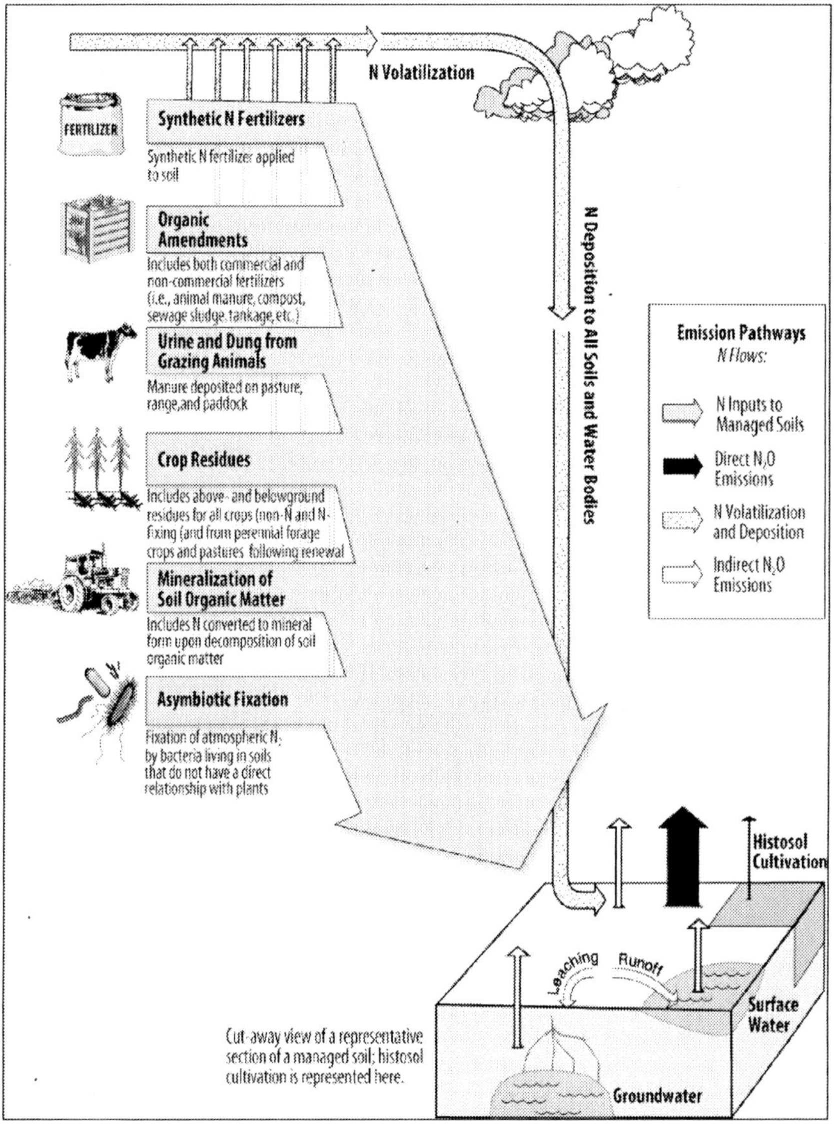

Source: U.S. Environmental Protection Agency, *2009 U.S. Greenhouse Gas Inventory Report*, EPA 430-R-09-004, Chapter 6, April 2009. Adapted by CRS.

Figure 1. Sources and Pathways of Nitrogen (N) Resulting in N_2O Emissions from Agricultural Soil Management.

The Nitrogen Cycle

Comprehension of the nitrogen cycle (**Figure 3**) is beneficial when crafting policy to reduce N_2O emissions from anthropogenic sources. Nitrogen, an essential element required by organisms to grow, is found throughout the atmosphere in various forms. The nitrogen cycle portrays the routes in which nitrogen moves through the soil and atmosphere in both organic and inorganic form. Certain processes within the nitrogen cycle convert the nitrogen into a form that can be taken up by plants. Four of the major processes are:

- nitrogen fixation—conversion of nitrogen gas (N_2) to a plant-available form;
- nitrogen mineralization—conversion of organic nitrogen to ammonia (NH_3);
- nitrification—conversion of ammonia (NH_3) to nitrate (NO_3-) via oxidation (that is, by being combined with oxygen); and
- denitrification—conversion of nitrates back to nitrogen gas.

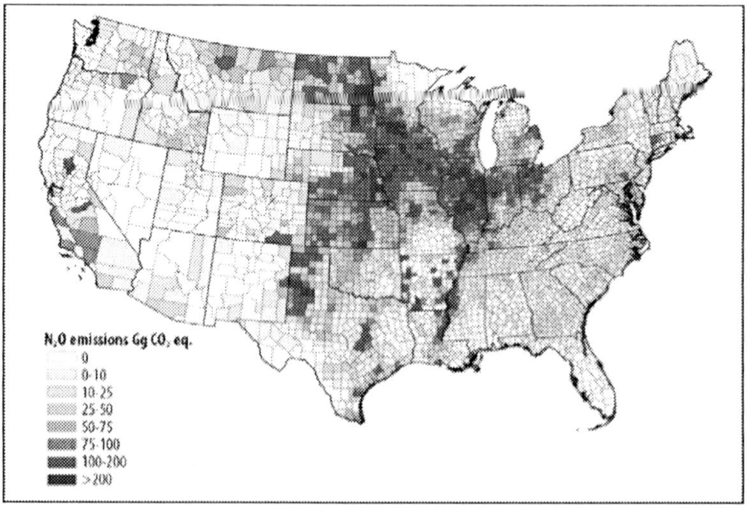

Source: USDA, U.S. Agriculture and Forestry Greenhouse Gas Inventory: 1990-2005. Adapted by CRS.

Notes: 1 Gigagram (Gg) is equivalent to 1,000 metric tons. Major crops are defined as corn, soybean, wheat, hay, sorghum, and cotton.

Figure 2. County-Level N_2O Emissions from Major Cropped Soils in 2005.

Nitrous oxide is a byproduct of nitrification and denitrification. Both processes occur naturally. Excess application of nitrogen fertilizer can lead to increased nitrification, which can cause nitrate to leach into groundwater or surface runoff (in turn, this causes eutrophication, which can damage aquatic environments).

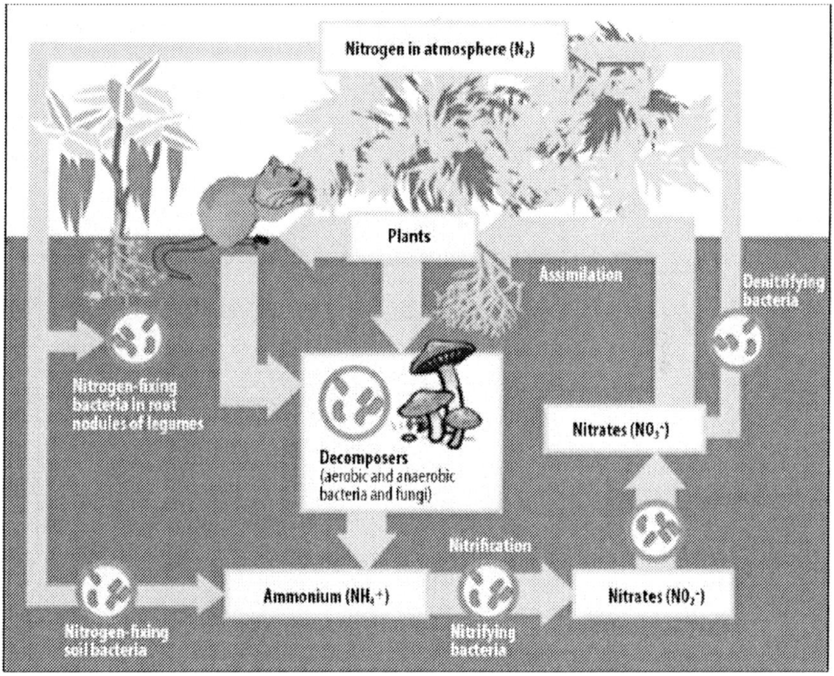

Source: EPA. Adapted by CRS.

Figure 3. The Nitrogen Cycle.

Opportunities and Challenges for Nitrous Oxide Emission Reduction

N_2O emission mitigation options are available for agricultural soil management and nitric acid production. Nitric acid is a chemical compound used to make synthetic fertilizers. N_2O abatement options for nitric acid production include a high-temperature catalytic reduction method, a low-temperature catalytic reduction method, and nonselective catalytic reduction.[13]

The estimated reduction efficiencies (the percentage reduction achieved with adoption of a mitigation option) are 90%, 95%, and 85%, respectively.

Agricultural soil management mitigation options recommended by researchers and technology transfer specialists to discourage excess application of nitrogen fertilizers and soil disturbance (**Table 2**) are not generally being practiced. Fertilizer and soil best-management practices aim to provide the crop with the nutrient and soil conditions necessary for crop production, and prevent nutrient and soil loss from the crop field (e.g., erosion, leaching).[14] Some may consider less money spent towards fertilizer use an economic incentive for agricultural producers.[15] Others may want to ensure that crop yields meet expected feed, fiber, and fuel mandates (e.g., for corn ethanol), which may be difficult to attain with less fertilizer use.[16]

Table 2. Select N_2O Mitigation Alternatives for Agricultural Soil Management

Mitigation Alternative	Description
Split fertilization	Application of same amount of nitrogen fertilizer as in baseline but divided into three smaller increments during crop uptake period to better match nitrogen application with crop demand and reduce nitrogen availability for leaching, nitrification, denitrification, and volatilization.
Simple fertilization reduction (10%, 20%, or 30%)	Reduction of nitrogen-based fertilizer from one-time baseline application of 10%, 20%, or 30%.
Nitrification inhibitor	Reduces conversion of ammonium to NO3-, which slows the immediate availability of nitrate (nitrate is water soluble). The inhibition of nitrification reduces nitrogen loss and increases overall plant uptake.
No-till	Conversion from conventional tillage to no-till, where soils are disturbed less and more crop residue is retained.

Source: EPA, *Global Mitigation of Non-CO_2 Greenhouse Gases*, http://www.epa.gov /climatechange/economics/ downloads/GM_SectionV_Agriculture.pdf.

Monitoring reduced nitrogen fertilization applications on a large scale for greenhouse gas emission reduction purposes may be difficult; it is not clear how such a program could be managed at a national level.[17] Enforcement options could include voluntary verification, third-party verifiers, or government intervention.

Reporting N_2O emissions from agricultural soil management was not included in the Final Mandatory Reporting of Greenhouse Gases Rule issued by EPA on September 22, 2009.[18] EPA's reasoning behind this decision was that no low-cost or simple direct N_2O measurement methods exist. Additionally, EPA released a proposed rule requiring new or modified facilities that could trigger Prevention of Significant Deterioration (PSD) permitting requirements to apply for a revision to their operating permits to incorporate the best available control technologies and energy efficiency measures to minimize GHG emissions.[19]

Federal Support for Nitrous Oxide Emission Reduction

USDA provides some financial and technical assistance for nutrient management through its conservation programs.[20] Moreover, USDA's Agricultural Research Service (ARS) is studying the relationship between agricultural management practices and nitrous oxide emissions.[21]

In addition to the agriculture sector, work is being done in the transportation sector to reduce N_2O emissions. Mobile combustion was responsible for roughly 8% of N_2O emissions reported in 2009.[22] One N_2O emission reduction effort, proposed by EPA and the Department of Transportation, is a per-vehicle N_2O emission standard of 0.010 grams per mile effective in model year 2012 for all light-duty cars and trucks as part of a wider effort to reduce greenhouse gas emissions and improve fuel economy in tandem.[23] EPA has allocated financial resources to quantify N_2O emissions for the greenhouse gas inventory (e.g., DAYCENT model).[24]

Policy Options for Nitrous Oxide Emission Reduction

Congress has begun to investigate the reduction of non- CO_2 greenhouse gas emissions, including N_2O emissions, as one strategy to mitigate climate change. Some contend that N_2O emissions reduction could serve as a short-term response in the larger, long-term scheme of mitigation and adaptation efforts.[25] It may be viewed as a short-term response because N_2O emissions make up a small amount of the GHG inventory compared to CO_2 emissions. Any substantial approach to mitigate climate change is likely at some point to have to address sources that emit CO_2.

Congress could approach N_2O emissions reduction as part of a comprehensive GHG emission strategy offering economically attractive abatement alternatives to discourage actions leading to climate change. For example, a cap or fee on N_2O emissions could spur innovative methods for agricultural producers to limit excess synthetic fertilizer application. Congress could also examine the tools necessary to identify N_2O emission abatement options, assess their cost, and determine their economic impact for full incorporation into climate change legislation.

Besides greenhouse gas emission reduction, reducing N_2O emissions could lead to ozone recovery. Congress could explore the co-benefits that may arise from restricting N_2O emissions for climate change purposes. N_2O is not regulated as an ODS under the Clean Air Act, Title VI, Stratospheric Ozone Protection (as guided by the Montreal Protocol). As emissions of other ODSs (e.g., chlorofluorocarbon-11, halon-1211) have declined due to regulation, N_2O has emerged as the dominant ODS emission.[26] The first-ever published ozone depletion potential (ODP) value assigned to N_2O, 0.017, is less than the ODP value of 1.0 for the reference gas chlorofluorocarbon 11 (CFC-11). While some may not see a cause for alarm based on the ODP value alone, the quantity of N_2O emissions and its potency as a GHG can lead to serious harm (see **Table 1**).

The ODP value for N_2O does not allow for its mandatory inclusion as a class I substance for regulation under the Clean Air Act.[27] However, N_2O could be listed as a class II substance at the direction of the EPA Administrator or regulated under Section 615 of the act.[28] Class I substances have an ODP of 0.2 or more and are more harmful to stratospheric ozone molecules than Class II substances, which have an ODP of less than 0.2.

With or without ODP substance listing, Congress may find it useful to incorporate the ozone depletion impacts of N_2O into its climate change policy proposals both to reduce greenhouse gas emissions and to further ozone recovery achievements. Classifying N_2O emission reduction as an eligible offset type, including N_2O as a covered entity within a cap-and-trade program, or directing EPA to use existing authority under the Clean Air Act to regulate N_2O are other available options to reduce N_2O emissions for ozone or climate protection. Any option chosen to reduce N_2O emissions will more than likely require an improvement of N_2O estimation, measurement, and reporting methods and possible financial incentives.

Congress could apply lessons learned from previous international agreements that are intended to abolish harmful compounds. The outcomes of the Montreal Protocol, put into action in the late 1980s, may prove useful to

Congress in understanding the long-term implications of certain climate change policy options, specifically cap-and-trade. A number of gases were phased out under the Protocol, which allowed for each country to establish a regulatory framework to monitor and reduce ODSs. Certain ozone-depleting substances, such as N_2O, were not included in the Protocol partly because their threat was not perceived as urgent at the time. However, one unintended consequence of the success of the Protocol reducing targeted ODSs is that N_2O has emerged as the leading ODS.

End Notes

[1] For more information on CO_2 emission reduction, see CRS Report RL33801, *Carbon Capture and Sequestration (CCS)*, by Peter Folger, and CRS Report RL34621, *Capturing CO_2 from Coal-Fired Power Plants: Challenges for a Comprehensive Strategy*, by Larry Parker and Peter Folger. For more information on legislative proposals to address climate change and regulation of greenhouse gases under the Clean Air Act, see CRS Report R40556, *Market-Based Greenhouse Gas Control: Selected Proposals in the 111th Congress*, by Larry Parker, Brent D. Yacobucci, and Jonathan L. Ramseur; and CRS Report R40585, *Climate Change: Potential Regulation of Stationary Greenhouse Gas Sources Under the Clean Air Act*, by Larry Parker and James E. McCarthy.

[2] U.S. Environmental Protection Agency, *2011 U.S. Greenhouse Gas Inventory Report*, EPA 430-R-11-005, April 2011, http://www.epa.gov/climatechange/emissions/usinventory report.html.

[3] The potency of a greenhouse gas is described by its global warming potential (GWP), an estimate of how much a greenhouse gas affects climate change over a quantity of time relative to CO_2, which has a GWP value of 1. Intergovernmental Panel on Climate Change, *Climate Change 2007: The Physical Science Basis* (2007), p. 212. The Intergovernmental Panel on Climate Change (IPCC) assigns nitrous oxide and methane a global warming potential of 310 and 21, respectively, over a 100-year time horizon.

[4] An ozone-depleting substance is a compound that contributes to stratospheric ozone depletion by releasing chlorine or bromine atoms into the atmosphere when broken down, leading to the destruction of ozone, a substance necessary to prevent harmful UVB rays from reaching Earth.

[5] The Montreal Protocol is an international treaty constructed to protect the stratospheric ozone layer by gradually eliminating a number of ozone-depleting substances.

[6] Also in the agriculture sector, animal digestive systems and manure management account for a large portion of U.S. methane emissions.

[7] Strategies and technologies for N_2O emission reduction are limited in comparison to resources expended for methane capture. Methane capture technologies, as well as financial and technical support, for point sources have been available for decades. For more information on methane capture, see CRS Report R40813, *Methane Capture: Options for Greenhouse Gas Emission Reduction*, by Kelsi Bracmort et al.

[8] S. Solomon, D. Qin, and M. Manning et al., *Climate Change 2007: The Physical Science Basis. Contribution of Working Group I to the Fourth*, IPCC, IPCC WG1 AR4 Report, New York, NY, 2007, http://ipcc-wg1.ucar.edu/wg1/ wg1-report.html.

[9] Agricultural soil management includes practices that add to, or create an environment conducive to the release of, mineral nitrogen (N).

[10] U.S. Environmental Protection Agency, *2010 U.S. Greenhouse Gas Inventory Report*, EPA 430-R-10-006, April 2010, http://www.epa.gov/climatechange/emissions/usinventory report.html.

[11] Mobile (fuel) combustion leads to N_2O being emitted as a byproduct. N_2O is released as a byproduct during the oxidation of ammonia for production of nitric acid, a primary component of synthetic fertilizers and some explosives. N_2O emissions are generally released in large amounts from dry manure handling systems (e.g., pasture, solid storage).

[12] U.S. Department of Agriculture, *U.S. Agriculture and Forestry Greenhouse Gas Inventory: 1990-2005* , Technical Bulletin No. 1921, 2008, http://www.usda.gov/oce/global_change /AFGGInventory1990_2005.htm. Nitrogen fixation is the conversion of nitrogen gas to ammonia.

[13] U.S. Environmental Protection Agency, *Global Mitigation of Non- CO_2 Greenhouse Gases*, EPA-430-06-005, 2006. Catalytic reduction methods use a catalyst to reduce nitrous oxides in exhaust gas at varying temperatures.

[14] C. S. Snyder, *Fertilizer Nitrogen BMPs to Limit Losses That Contribute to Global Warming* , International Plant Nutrition Institute, Ref. # 08057, June 2008, http://www.ipni.net/ipniweb /portal.nsf/0/6D54ABC2C92D9AFA8525749B0074FF59.

[15] According to the Government Accountability Office, natural gas is the highest-priced factor when producing nitrogen fertilizer. Thus, natural gas prices impact nitrogen fertilizer costs. U.S. General Accounting Office, *Natural Gas: Domestic Nitrogen Fertilizer Production Depends on Natural Gas Availability and Prices*, GAO-03-1148, September 2003, http://www.gao.gov/new.items/d031148.pdf

[16] For example, the Renewable Fuel Standard (RFS) is a provision established by the Energy Policy Act of 2005 requiring gasoline to contain a minimum amount of fuel produced from renewable biomass (including corn). For more information on the RFS, see CRS Report R40155, *Renewable Fuel Standard (RFS): Overview and Issues*, by Randy Schnepf and Brent D. Yacobucci.

[17] For more information on monitoring carbon in agriculture, see CRS Report RS22964, *Measuring and Monitoring Carbon in the Agricultural and Forestry Sectors*, by Ross W. Gorte and Renée Johnson.

[18] For more information on the agricultural implications of the Mandatory Reporting of Greenhouse Gases Rule, see CRS Report RL32948, *Air Quality Issues and Animal Agriculture: A Primer*, by Claudia Copeland.

[19] For more information on the proposed PSD rule, see CRS Report R40585, *Climate Change: Potential Regulation of Stationary Greenhouse Gas Sources Under the Clean Air Act*, by Larry Parker and James E. McCarthy; and EPA, *Proposed Rule: Prevention of Significant Deterioration and Title V Greenhouse Gas Tailoring Rule*, 2009, http://www.epa.gov /NSR/fs20090930action.html.

[20] For more information on agricultural conservation programs, see CRS Report R40763, *Agricultural Conservation: A Guide to Programs*, by Megan Stubbs. For more information on technical assistance for nutrient management, see CRS Report RL34069, *Technical Assistance for Agriculture Conservation*, by Megan Stubbs.

[21] For more information on the efforts underway at ARS, visit the Air Quality of Agricultural Systems Research Unit website at http://www.ars.usda.gov/main/site_main.htm?modecode=36-25-15-15, or the Air Quality National Program website at http://www.ars.usda.gov/research/programs/programs.htm ?NP_CODE=203.

[22] U.S. Environmental Protection Agency, *2011 U.S. Greenhouse Gas Inventory Report*, EPA 430-R-11-005, April 2010, http://www.epa.gov/climatechange/emissions /usinventory report.html.

[23] Environmental Protection Agency, Department of Transportation, "Light-Duty Vehicle Greenhouse Gas Emission Standards and Corporate Average Fuel Economy Standards; Final Rule," 75 *Federal Register*, May 7, 2010.

[24] EPA uses the DAYCENT ecosystem model for the U.S. greenhouse gas inventory "to estimate direct N_2O emissions from mineral cropland soils that are managed for production of major crops—specifically corn, soybeans, wheat, alfalfa hay, other hay, sorghum, and cotton."

[25] Shilpa Rao and Keywan Riahi, "The Role of Non- CO_2 Greenhouse Gases in Climate Change Mitigation: Long-Term Scenarios for the 21st Century," *Energy Journal*, vol. 27 (2006), pp. 1-26; Mario Molina, Durwood Zaelke, and K. Madhava Sarma et al., "Reducing Abrupt Climate Change Risk Using the Montreal Protocol and Other Regulatory Actions to Complement Cuts in CO_2 Emissions," *Proceedings of the National Academy of Sciences of the United States of America*, October 12, 2009..

[26] A. R. Ravishankara, John S. Daniel, and Robert W. Portmann, "Nitrous Oxide (N_2O): The Dominant Ozone-Depleting Substance Emitted in the 21st Century," *Science Express*, August 27, 2009.

[27] 42 U.S.C. §7671a. The EPA Administrator may add to the list of class I substances any substances that the Administrator determines as having an ozone depletion potential of 0.2 or greater.

[28] 42 U.S.C. §7671n. The EPA Administrator has the authority to promptly promulgate regulations respecting the control of an ODS by submitting notice of the proposal and promulgation of such regulation to the Congress.

INDEX

D

E

F

G

Q

R

S

T

U

V

W

Y